D1458306

The Private Life of Plants

BY THE SAME AUTHOR

Shrubs and Decorative Evergreens
Trees for Smaller Gardens
Ferns
Grow Your Own Vegetables
Bottle Gardens
The Two Hour Garden
The Natural Garden
Growing Vegetables and Herbs
Your Greenhouse, How to Grow and Save
Fresh From the Garden (with Robin Howe)
Ornamental Grasses

The Private Life of PLANTS

Roger Grounds

Illustrations by Meg Rutherford

DAVIS-POYNTER
LONDON

First published in 1980 by
Davis-Poynter Limited
20 Garrick Street London WC2E 9BJ

ISBN 0 7067 0216 6

Photoset by Photobooks (Bristol) Ltd 28/30 Midland Road St Philips Bristol
Printed in Great Britain by The Stellar Press Hatfield Hertfordshire

Contents

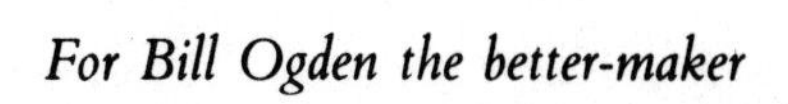

For Bill Ogden the better-maker

PART ONE
FOREPLAY

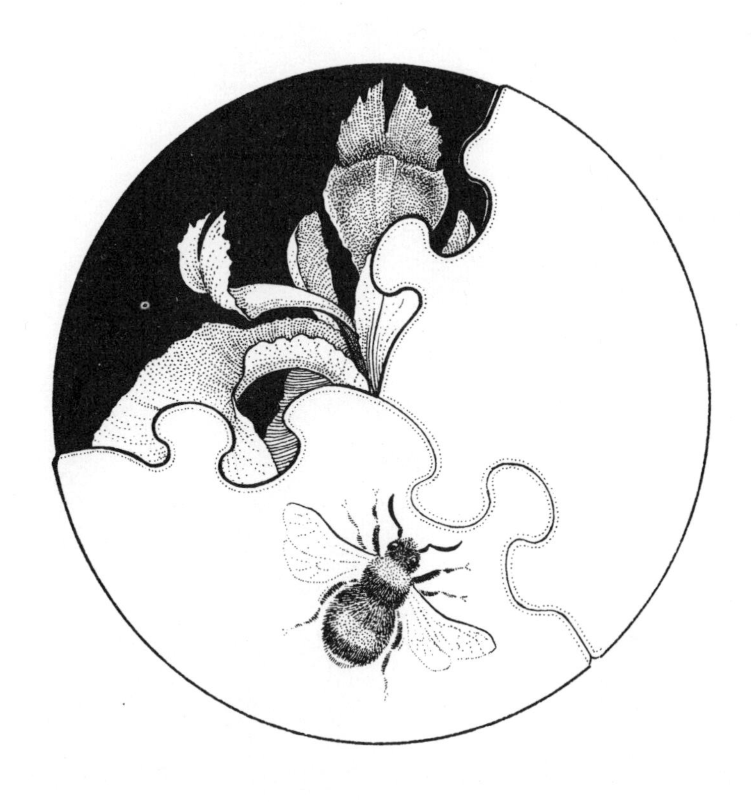

1
Eccentric and Bizarre

THE SEX LIFE of plants is so fascinating, bizarre and at times comic that it seems incredible that, in an age when the sex lives of everything else from ants to apes has been both scrutinized and publicized, the sex life of plants should remain shrouded in the decent obscurity of a learned language, comprehensible only to those who suffer the encumbrance of a previous education.

The sex lives of men and monkeys may amuse, but they dull beside the sheer exuberant sexuality of the plant kingdom and the brilliant diversity of its sexual inventiveness. The most that man (embracing woman) can accomplish with his limited equipment is a small number of variations on a handful of basic positions. Plants with their greater diversity of sexual equipment, arranged in amazing permutations of patterns, can achieve far, far more – so much indeed that the most extravagant excesses of the most prurient pornographers seem tame by comparison. What might seem the most vile perversions in man, may be normal among plants. So often what is normal among animals, is exceptional among plants, and the other way about. So that cunnilingus is as much a normal part of the sex life of most plants as copulation is of most animals.

Perhaps the essential difference between people and plants is that whereas among people the sexual organs are very much the private parts, among plants they are the public parts. For flowers are nothing less than the sexual organs of plants, most blatantly displayed.

If all this is not always apparent this is at least partly because there seems to have been an almost deliberate attempt to conceal this simple truth from us in an unfamiliar terminology. Gone are the tender terms we use when we speak to one another, lover to lover of our sexual organs: there is not even the technical terminology so familiar from text-books. Instead we meet words we never met before, words that are cold, harsh, lacking in emotive human overtones. The male organ of a flower is not a penis, but a flimsy-sounding thing called a filament, while the sensitive, bell-shaped glans has been disfigured into an anther. The female organ is not the familiar vagina, but a thing called a style (as though it were some obstacle to be overcome). By a curious quirk the male organs all sound female, and the female organ sounds male. A style certainly sounds more virile and vigorous than a filament, which sounds so fragile one suspects that it might snap off if it were to try to partake of any sexual activity. No doubt if plants had testicles the terminology would have castrated them too.

Such terminology is, essentially, Victorian. Its purpose was not so much to denature the sexuality of plants, as that of people. Theirs was, after all, an age in which young ladies of gentle birth were not allowed to wash between their legs for fear lest this arouse in them the sin of pleasurable sensations. In using this remote and arid terminology, they have alienated us from the plants with which we share this planet and upon which we are so dependent, leaving us estranged.

How much more interesting it would all be if, with the first stirrings of our own sexuality, we were taught about plants in such a way that we could relate what happens in humans to what happens in plants. If we were told, for example, that each of the pubic silks tufted round a corncob was an individual vagina, moist and ready to suck up the pollen-sperm brought to it on the wind, and that the pollen grain, a sort of wandering wind-blown penis, would have to grow and penetrate the vaginal silk, (penetrating it right to the very bottom), where the sperm could fertilize the waiting ovule, and that each kernel on a corn cob is an individual ovule, and that each of the multitude of seeds on a corn cob is the result of an individual act of copulation.

How much better we would understand plants if we knew that the ovule is to a plant what a womb is to a woman. That in most (but not all) plants each womb has only one ovule or egg to be fertilized and, as in humans, it takes one sperm to fertilize each ovule or egg cell. That there are, on average, about 2,500 seeds in the capsule of a tobacco plant. And each one of those seeds took an individual sperm to fertilize it – 2,500 separate acts of copulation. And all that took place in less than twenty-four hours in a space less than 1 mm across.

If the sexuality of plants were made explicit in this way there would be a great many people who would have a far better understanding in human terms of what to do with which and when than they do now: it is difficult to relate to an amoeba, which has no sexual parts, or to a frog which has no penis: plants, in many ways, are far more like people.

Far more like people than most of us realize. Flowers, like women, emit a powerful and seductive odour when ready for mating. If they remain unmated they continue to emit this odour until they fall, perhaps seven or eight days later. After mating and fertilization, the mating odour ceases, usually within less than half an hour of fecundation taking place. In flowers, as in many animals (including women) a different odour is frequently emitted once mating has successfully taken place, an odour which warns approaching lovers not to waste their time; birds and bees, unlike humans, find life too short to waste time mating with a female that has already been mated. In humans sexual frustration can express itself in a variety of ways, most of them unpleasant: one of the most common ways is by the emission of a strongly repulsive smell. The same is true of flowers. In flowers, as in females, human and other, there is a measurable increase in temperature in the female organ when it is ready for mating: flowers, like animals, quite literally come on heat.

The point is seldom ever made that hermaphroditism is as common in plants as it is rare in animals. In most flowers it is very obvious that both filament (penis) and stigma (vagina) are present. In many flowers such as the primrose, *Primula vulgaris*, while both organs are present, one is more developed than the other. The thrum-eyed and pin-eyed primroses are not different species as used to be thought. It is simply that in the thrum-eyed primrose the anthers (male organs) are visible, the stigma (vulva) being hidden, while in the pin-eyed primrose the stigma is plainly visible, the male organs being reduced. A similar situation, in which male and female are distinguished from each other occurs also in man. The clitoris is, in essence, a much reduced penis, while the glans is a modified labiate vulva.

Amazing theatrical effects can be obtained with pollen, the semen of plants. It is highly inflammable and, if thrown onto a red hot surface, will flare like magnesium. In the old days the pollen of clubmosses, *Lycopedium spp.*, was thrown onto a hot shovel hidden in the wings of the theatre. It is an experiment which can readily be tried provided only that you can recognise a clubmoss when you see one.

Much more pertinent is the point that the pollen of many plants emits a strong odour quite remarkably similar that of the semen of many animals, including man. Further similarities exist between pollen and semen, apart of course from the fact that they both serve the same function. The pollen, once it lands on the stigma (vulva) immediately begins to grow a pollen tube which, like the erect penis, has to penetrate the style (vagina) before releasing the sperm.

In many plants, as in many animals and in humans, much sexual activity is governed by smell. The sperms of many mosses are guided in their search for a female ovule by a strong chemical scent; malic acid, in this case. The sperms of ferns have a sweet tooth, and are guided in their search for an ovule by a sweet, sugar-like smell. The females of many animal species have special scent secreting glands by means of which they can mark their territory and so attract a mate.

It may seem obvious today that plants reproduce sexually, yet the discovery of sex in plants is a relatively recent one – about as recent as the invention of the railway train.

The first person to stumble on some sort of sexuality in plants was the Greek, Theophrastus. He describes in his book *On the History of Plants* how 'With the date it is helpful to bring the male to the female; for it is the male which causes the fruit to ripen and to persist . . . The process is thus performed; when the male palm is in flower, they at once cut off the spathe on which the flower is, just as it is, and shake the bloom with the flower and the dust over the fruit of the female, and, if this is done to it, it retains the fruit and does not shed it. In the case of both the fig and the date it appears that the 'male' renders aid to the 'female'—but while in the latter case there is a union of the sexes, in the former the result is brought about somewhat differently,' (and thereby hangs a very strange tale).

Sadly Theophrastus had not grasped plant sexuality at all. It seems that he and his contemporaries used purely anthropomorphic qualities for judging male from female, the male being more masculine in appearance, the female more delicate, more elegant.

No further progress was made until the development of the magnifying glass and the earliest microscopes. With a magnifying glass it was suddenly possible to see details in flowers which until then had been too intricate for the human eye to perceive. Once the intimate organs of flowers could be seen it did not take long for their true function to be realized.

It was the English botanist Nehemiah Grew who first put into print the suggestion that there might be some phallic implication in the 'attire' of the flowers: he said that he had discussed the connexion of the stamens and the formation of seed with Thomas Millington (at one time Sedlian Professor of Natural History at Oxford) who had suggested to him 'that the attire [stamens] doth serve, as the male, for the generation of seed . . .' a proposition with which Grew agreed.

That greatest of English botanists, John Ray, whose massive *Historia Plantarum* is such a landmark in botanical literature, was not wholly convinced: he said 'This opinion of Grew, however, of the use of pollen before mentioned, wants yet more decided proofs . . .'

Such more decided proofs were not long in coming. Rudolph Jakob Camerarius, professor of Physic at Tubingen in Germany carried out experiments as a result of which he was able to conclude 'Hence it appears wholly reasonable to assign to the apicés [stamens] themselves a nobler name and the function of the genital parts of the male sex, as their capsules are vessels and containers, in which the semen itself, that powder, the most subtle part of the plant, is produced, collected and from here afterwards given out . . . Plants exhibit equally these apices as the factory of the male semen and the seed vessel with its little feather or style as the genital parts proper to the feminine sex.'

In spite of which Joseph Pitton de Tournefort, a capable enough botanist, was able to state that the stamens were to a flower, very much what an anus is to an animal, the pollen merely being the waste matter excreted by the plant in powder form.

Yet Tournefort described some 8,000 species of plants belonging to twenty-two classes, and so set the stage for the progenitor of modern botanical classification, Carl Linné, better known as Linnaeus.

Linnaeus is usually considered the 'father' of modern botany, as though he had begotten the whole plant kingdom himself. (No one seems to know who the 'mother' of modern botany was.) In fact what Linnaeus achieved was a marriage of two streams of thinking, a marriage so effective that it took over 100 years for the divorce to be effected. Linnaeus' true passion was not plants: it was order. His fundamental belief seems to have been that God created an orderly world, and that his job was to find that order.

The first partner in Linnaeus' marriage was the realization that the flowers of plants are their sexual organs displayed in full frontal nudity. His frankness certainly went further than that of anyone before him. He realized that a more precise method of classifying plants could be derived from a study of their sex organs than could have been created from any previous system. He divided the plant kingdom into species and genera according to the variations of the stamen (male organ).

The great advantage of this so-called 'sexual system' was that it could be used by anyone able to count up to twenty. It therefore caught on very quickly. All you had to do was count the number of stamens noting where necessary whether the filaments were united or not, and this led you in due course to the appropriate order, which was further subdivided by characteristics anyone could recognize, into genera and species. Thus if the plant before you had nine stamens and six anthers, you could fairly readily work out that it was a flowering rush.

The other partner in Linnaeus' marriage was the naming of plants. Before his time a multi-nominal system had been in use. This was often confusing. The multi-nominals were basically short descriptions. Thus *Gramen caninum aristatum radice non repente sylvaticum* (The coarse awned woodland grass with a non-creeping rootstock) was the fibrous couch grass, and not too bad a description of it at that, since it sufficiently distinguishes it from the other couch grass, the running one, *Agropyron repens*. Multi-nominals were not always as helpful as this: what for example, is the plant described as *Erica pumila calyculata unedois flore* (The little heath with the small calyx and a flower like a strawberry tree)? A heath – but which? Similarly, which of the multitude of pinks is *Dianthus silvestris altera calicula foliolis fastigiatis* (The other wild pink, with a little calyx and erect leaves)?

In place of this system Linnaeus proposed a system in which each plant was assigned only two names. The first was the generic name (the name of the general group to which the plant belonged), the second was the specific name (the name by which the plant could be specifically identified from others in the same group). Thus the name *Typha* is the general or generic name for the reedmaces or cattails: *T. latifolia*, is the broad-leaved cattail, the *latifolia* being the name of the species or the way in which that particular cattail specifically differs from others, such as *T. angustifolia* the narrow-leaved cattail.

This system is so handy that we still use it today. It is what is known as the binominal system. However, when Linnaeus launched his *Systema Natura* upon an unsuspecting world the two systems were so married that it was nearly a century before people realized that the sexual system was one thing, and the binominal system something quite different.

Linneaus pompously entombed his masterwork in Latin, even at that time a language little read. Had it remained in that language it would probably not have created the sensation it did. It was its translation into English which caused the real furore. It was decided in place of Linnaeus' forthright terminology to use a mode of expression more acceptable to a wide public. Thus the corolla became the marriage bed, the stamens the husbands and the styles the wives. With some curious results. Thus the class of plants Linnaeus named *Monandria* was translated as 'one husband in a marriage', while *Driandria* became 'two husbands in the same marriage.' So far so good, but some of his other classes sounded in translation, as though they had come straight from a Roman orgy. *Polyandria* is explained as 'twenty males or more in the same bed with the female', while *Syngenesia Polygamia Necessaria* was 'Confederate males with Necessary Polygamy'. This class included, for example, the marigold, with its fertile ray and sterile disc florets, which in translation read: 'the beds of the married occupy the disc, those of the concubines the circumference; the married females are barren, the concubines fertile.' One cannot help wondering what a modern psychologist would have made of that, it seeming more natural that the married females should be fertile. But worse was to follow: there were curious and unnatural phenomena such as 'husbands joined together at the top.'

Incredible though it may seem, this translation was not a joke: it was a perfectly serious translation made by 'a Botanical Society of Lichfield' – one of the three authors being Erasmus Darwin, grandfather of Charles Darwin. The translation came in for horrified criticism, especially from the Church. The Rev. Samuel Goodenough, later Bishop of Carlisle, wrote, in a letter: 'To tell you that nothing could equal the gross prurience of Linnaeus' mind is perfectly needless.' While Goethe worried himself silly about what effect the botanical textbooks of the future might have on the tender minds and morals of schoolchildren of the future.

Erasmus Darwin was to go on to do even worse: he versified the Linnaean system in a poem entitled *The Loves of the Plants*, the coyness of which is far more prurient than the frank forthrightness of Linnaeus himself. He describes the three stigmas and six stamens of 'fair Colchica' as:

> Three blushing Maids the intrepid Nymph attend,
> And six gay youths, enamour'd train! defend . . .

(It may just be worth mentioning that in those days the word 'gay' did not carry the connotations attached to it today.) His description of the glory lily, in which three of the six stamens mature later than the other three, is so coy that it is far, far worse than a blunt statement of the truth:

> Proud Gloriosa led three chosen swains,
> The blushing captives of her virgin chains.
> When time's rude hand a bark of wrinkles spread
> Round her weak limbs, and silver'd o'er her head,
> Three other youths her riper years engage,
> The flatter'd victims of her wily age.

a pathetic fallacy which makes the female of the species sound a great deal more revolting than La Belle Haeulmiere and Villon's vision of old age. When we get to Silene we find that matters have deteriorated even further:

> The harlot-band ten lofty bravoes screen

No wonder Goethe had reason to fear for the morals of the coming generation!

It took a man of the genius of Charles Darwin to put this bawdy house in order. To Darwin the purpose of classifying plants was to try to show the evolutionary relationships between them. He saw, with penetrating clearness, that without sex there could be no evolution. It is only through the transmission of genes in the sexual process that change becomes possible, those changes which fit plants and animals best for survival being beneficial, hence the doctrine of the survival of the fittest.

The first specialized organs produced by any plants were sex organs. From then on, they seem to have put more effort into refining their sex organs than into any other part of their anatomy. Indeed, the main purpose in the lives of plants at all levels is to reproduce. All their other life processes are subservient to this end.

By human standards, the sex life of plants is decidedly weird. There are flowers that quite literally come on heat like animals; flowers that mate only in the dark, flowers that copulate underground, flowers that make the insects which come to pollinate them drunk, others that kill them and some so closely resemble the female of the species which comes to pollinate it that the male goes through a kind of pseudo-copulation with it.

And though all this was known long, long ago, the Victorians were still able to believe that the bee orchid looks like a bee to frighten other insects away, and Tennyson was able to write, in spite of all the evidence to the contrary, of 'The wild flower of blameless life'!

2
Of Man and Plant

TO MOST PEOPLE it seems perfectly obvious that plants are, in some way not always easy to define, radically different from us. Or are they?

The only workable differentiation is that plants possess chlorophyll, which animals do not. Curiously, this is not such a substantial difference as it may seem.

Chlorophyll is essentially a pigment. It absorbs the radiant energy from sunlight and uses this to weld six molecules of carbon dioxide and six molecules of water into one molecule of glucose, in the process discarding six molecules of oxygen. The subtances that chlorophyll welds together form food for plants, and for the animals that feed on plants, while the oxygen molecules the chlorophyll discards form the very air we breathe.

The interesting thing about chlorophyll is that part of its molecular structure is almost identical with that of haemoglobin (also a pigment) in human blood. The main difference is that it is iron in blood which gives it its red colouring, while it is magnesium that gives chlorophyll its green colouring. Haemoglobin possesses the power to combine with oxygen, (which increases its redness) while chlorophyll possesses the power to combine with carbon dioxide.

Another interesting link between the plant and animal kingdoms is a substance called haematoxylin (used as a dye-stuff) which is obtained from the tree *Haematoxylon campeachianum*. What the tree yields is a nearly colourless crystalline substance that only becomes red when combined with oxygen. This red substance is called haematin. Haemoglobin from human blood is readily decomposed into the pigment haematin. The chemical structure of haematin from both sources is identical.

The similarities do not end there. In all higher plants the chlorophyll is contained in variously shaped bodies called chloroplasts, and there is a striking similarity between these and the rod cells in the eyes of vertebrates. Further, the parts of the visible light spectrum absorbed by the chloroplasts fall clearly within the range of the human eye, the orange, red and blue rays of the visible spectrum. Both rod cells and chloroplasts are almost identical in structure, each being less than a micron across and looking, as one writer has put it, like a stack of several hundred hollow balls squashed flat, piled up and then wrapped in a single, very thin, clear membrane. Both absorb the energy from light, but use it differently. The rod cells of the human eye act as a combination of amplifier and transducer, converting the energy absorbed from the light into nerve signals. The chloroplasts act as a semi-conductor, converting the radiant energy from sunlight into chemical energy. In both cases extremely rapid photochemical reactions are involved.

One intriguing thing about chloroplasts, a thing which few people seem to know, is that the energy can flow either way: while the normal mode of operation is for the chloroplasts to absorb light, at the moment when light gives way to darkness, the chloroplasts can actually emit light, albeit extremely feebly and for the briefest period. It is also thought, though not experimentally proven beyond all doubt, that some humans are capable of emitting energy through their eyes, and that this is an important factor in photokinesis.

In both plants and animals there are elaborate devices for preventing the light-absorbing cells from being damaged by too high an intensity of light. In most animals this is achieved by means of a variable aperture, such as the iris of the eye. In plants a number of devices have been utilised: many plants have hairs over the leaves, or scales which effectively prevent too much light reaching the chloroplasts, while in other plants the leaves can droop or fold, thus reducing the area of the leaf surface exposed to the sun and so effectively reducing the amount of light reaching the chloroplasts.

There are also far greater similarities in the way that plants and humans use their foodstuffs than is generally appreciated. What a plant actually produces as a result of combining carbon dioxide and water is glucose. Any glucose not needed by plants for immediate use is then converted by a chain of complex chemical reactions into 38 molecules of adenosine triphosphate (ATP for short). ATP is, as it were, the universal currency of energy in both plants and animals. By storing ATP not needed for immediate use in the form of carbohydrates and other complex organic sugar derivatives, the plant builds up, as it were, a store of capital assets in the form of fats, proteins and so on.

When we, as humans, eat a plant we convert its substance back into glucose, for immediate use, the rest going into storage in the form of carbohydrates, proteins and fats. Thus we firstly break down the complex substances which the plant has built up for its own use, and then duplicate the chemical reactions to turn it into forms in which it can be stored for slow release.

Plants, like people, sleep at night. A plant cannot photosynthesise at night, since there is no sunlight available. What it does is switch over to consuming oxygen and giving off carbon dioxide. The oxygen it consumes in this way enables it to 'burn' the foodstuffs it has stored during the day. It is by doing this that it builds up its body tissues, cells increase, stems thicken, reproductive, organs move towards maturity. Thus a plant does most of its growing and body-repair work by night. It is also at night that most growth in humans occurs, the reserves stored during the day being used to repair damaged cells and replace old cells.

The sleeping patterns of plants are quite remarkably similar to our own. Both follow what is known as the circadian rhythm – the word circadian meaning 'about a day.' It seems that, like us, plants have a built-in master clock which governs not only their sleeping and waking but which also, in computer jargon, handles a number of sub-routines. Very few humans have an inner clock that works to exactly 24 hours to the day; in some people the clock works to a shorter day, in some to a longer day: like a watch that runs fast or slow, our rhythms need 're-setting' from time to time, and most of us feel at our lowest performance when the clock has lost or gained too much, and though we may not realize it, we sleep for longer periods or lesser periods until the inner clock is in alignment with the day of the outer world. We are at our most energetic when both inner and outer clocks are keeping the same time.

Plants, like people, really do sleep, and wake: this is not mere anthropomorphism. Many plants either droop or fold their leaves at night, returning to the erect position on waking. Perhaps the most familiar plant in which this phenomenon can be readily observed is the so-called prayer plant, *Maranta leuconeura*, which folds its leaves together along the mid-rib at night, very much in the manner of hands at prayer.

The circadian rhythms of alternating periods of sleeping and waking are of fundamental importance to all living things and it is the arrival of dawn that enables the inner clock to be reset. This is particularly obvious in plants, where only with the coming of the sunlight can the photosynthetic processes start up again. However, waking is a far more complex matter than most of us realize. At whatever stage of evolution a creature has reached, it will have incorporated into itself something of every stage of evolution preceding it. The phenomenon is readily observed in the human foetus, which passes through every stage of evolution in the womb, from single-celled organism through fish (complete with gills) to mammal, to primate. Each of these different segments of the human, or of any plant, evolved at different periods, the geological time-span being enormous. In different geological periods the days were of different lengths, so that the circadian clock has to act as a master clock controlling each of the sub-routines which people and plants have incorporated.

The fungi, for example, are extremely primitive plants. Their circadian rhythm is, rather surprisingly, 4 days long, suggesting that they probably evolved at a period when the day was very much longer, or possibly evolved without clear circadian rhythm and are still groping their way towards one. Recent geophysical research into a group of fossils known as Stromatolites, a group which exhibits daily growth rings, suggests that in their time, about two and a half million years ago, the day was only five hours, while the lunar month was forty to forty-five days long. Fossil corals reveal that in Cambrian times, about six hundred million years ago, the year was some 425 days long, which implies a day-length shorter than the twenty-four-hour day as we know it. Even today we know that our day length is not stable, and that it is slipping by milliseconds in every twenty-four hours. Most plants have carried with them to the present the time-patterns imposed on them in the era when they evolved.

The circadian rhythm has therefore to handle a number of sub-routines which have an inherent tendency to get out of step. It is necessary to 'gate' these sub-routines so that they all start together at some particular point in time. In plants it is the arrival of daylight that starts all the routines running together: as the days go by, some of them lag behind, others complete their routine too fast: a period of rest is necessary for them all to stop, and to start again together. By modern computer standards a gate a whole night long is highly inefficient, but plants live their lives at a slower pace.

If the circadian rhythm in humans is upset they get extremely weary, nervy and irritable. If attempts are made deliberately to upset the natural clock by methods such as never allowing a person to sleep or by keeping them in a wholly artificial environment in which the day and night sequence can be altered at random, the whole mental, nervous and physical system simply breaks down because the sub-routines are all out of alignment. Plants too can be brought into a condition that could only be described as a nervous breakdown by similiar methods.

There are further similarities between plants and animals in the basic body chemistry of each. The most obvious of the chemicals shared in common are growth promoting and growth controlling substances. In plants the chemicals are auxins: in humans and other animals they are hormones. The chemical make-up of both bear a number of very close relationships. Some of each are found in both animals and plants. Some of the auxins usually associated with plants are found in human urine, for example. Further, the range of vitamins needed by plants is very similar to that needed by humans. Vitamins are also growth regulators, but different in kind from auxins and hormones.

One hormone which has caused a lot of interest recently and whose role in plants and animals is not yet fully understood is kinetin. It is found in both plants and animals, and kinetin derived from either source is the only substance known to produce growth in fragments of plant tissues. It is found in very high concentrations in coconut milk, and is believed to exist in a more diffuse way in all higher plants. It is also found in yeast, and in fish sperm.

Riboflavin is one of the most fascinating substances common to both man and plants. It is known to be intimately linked with vision. Though it is present in most plants in considerable quantities, its role in the life of plants has not been precisely determined. However, since plants are able to orient their leaves towards the sun it seems reasonable to hypothesise the presence of a light sensitive chemical in the plants, and riboflavin seems to be the likeliest candidate. It is possible that carotenoids, also essential to animal vision, may play a part here.

There are other phenomena shared by plants and animals, though many of these are as yet very little understood. It is known, for example, that plants have a pulse. This pulse reveals itself as a regular voltage fluctuation similar to that produced by the heart. In plants these voltage fluctuations occur at intervals of between one and ten seconds. Neither their origin nor their function has yet been properly explained.

Oestrogens occur in many long-day plants, and the quantities of oestrogen present build up as flowering approaches. For a very long time it was thought that oestrogen was an exclusively animal sex hormone: in fact it appears to be an almost universal sex hormone. If applied to certain plants it will induce them to rush into flowering far more rapidly than they would if left to their own devices. In long-day plants this female sex hormone builds up quite naturally if the plant is growing under ideal conditions. However, if grown in unsuitable conditions, such as, for example, being kept under perpetual short-day conditions, the hormone does not develop at all. Similarly in most animals oestrogen only builds up when conditions are right, when the animal is both healthy and happy.

Interesting though it is to find that plants and man have so much in common, it is all too easy to overlook the greatest point of common interest, the fact that both reproduce sexually, and that both have sex organs with a great deal in common.

Thus when we pause to take a really close look at plants it turns out that, far from them being radically different from us, they have far more in common with us than most of us ever realized. Indeed, as someone once glibly said, plants are only human, not least of all in the way they conduct their sex lives.

3
Can Plants Feel?

PLANTS, as we have already seen, are far more like people than most people realize. They have sap, we have blood; they have chlorophyll, we have haemoglobin; they have chloroplasts, we have eyes; both have a pulse; both have circadian rhythms; both produce hormones which in many cases are very similar; and both reproduce by means of sexual processes which are comparable in their complexity. Both give birth, their offspring go through a period of childhood (often marked by clearly juvenile characteristics), through puberty to maturity, mate, show obvious signs of ageing and, in time, die.

Whether or not we are consciously aware of the things we share with plants, we tend to attribute to them many other essentially human characteristics. When, for example, we see a rhododendron curl its leaves in cold weather in winter we naturally tend to assume that it is feeling the cold. When we see plants hang their leaves during hot, dry weather, we tend to assume that, like us, they feel the heat. When we observe flowers which open when the sun is shining, but close when a cloud passes by, we tend to assume that like us, flowers enjoy basking in the sun.

But these are only projections of our own human feelings. It is possible that their reactions may be of a purely mechanical kind. And if plants do feel at all, it is quite possible that they have a range of feelings quite different from our own. The question of whether or not plants can feel is an intriguing one; the answers may be even more intriguing.

In order to know what we are looking for we need to have an accurate and workable definition of what we mean by 'to feel'. The dictionary definition is 'to perceive or be aware of through physical sensation' (Webster). It is a simple definition, avoiding those confusions which so often arise over emotions and intellectual stimuli.

The lowest level of life at which it is an observable fact that feeling (as defined) takes place occurs in some very simple, single-celled organisms. The unicellular phytodiatoms, which live in water, move upwards to the surface during hours of daylight to position themselves so that their chloroplasts can use the energy from sunlight to photosynthesise. At night, they sink to the bottom to rest. Such deliberate movement is not merely mechanically activated. The plants do not move until they feel the sunlight: when they no longer feel the sunlight they rest. This is a very simple plant making a simple response to a simple stimulus.

The single-celled phytodiatoms are of more interest than they seem at first, both on account of their shape and the way they move. Each is shaped like a sperm. It has a head and a tail. At the centre of the head is a diffuse chloroplast, and above it an eye-patch, as it were: the phytodiatom moves itself in relation to the sun so that the eye-patch covers the light-sensitive cell for resting periods. Were it not able in this way to protect its light-sensitive area the energy from sunlight would quickly burn it out. The tail of the phytodiatom is used as is the tail of a tadpole or a sperm, to propel it by a wiggling, whip-like motion. The male sex cells of slightly higher plants are once again tadpole-shaped. Even in the highest animals, the male reproductive cell still has this very elementary tadpole shape.

The implication of this, of course, is that if a phytodiatom can feel, then so too can a sperm. Whereas what a phytodiatom feels is light, what the sperm of a moss feels is the attraction of a chemical scent emitted by the female organ, and so moves towards it. In this case the scent is malic acid. The sperm of a fern feels the attraction of a sweet, sugar-like chemical 'scent'. The sperms of higher plants and animals are similarly guided. Taste is as much a feeling or physical sensation as touch or sight.

Since it is demonstrable that a single cell can feel it would seem logical to assume that the more complex an organic structure the greater its ability to feel. Since we are unable to communicate directly with the plants (at least at present) the only way in which we can establish whether or not a plant feels a particular stimulus is according to whether or not it reacts to it. The reaction we are looking for is movement.

There is in fact a large corpus of documentation showing that plants move in response to stimuli, or at least that particular parts of plants move in relation to stimuli that are relevant to their function. For example, just as phytodiatoms move upwards towards the light, so the leaves of most plants turn themselves towards the source of greatest light intensity, and because they do this the whole plant tends to grow towards the source of the light. The point here is that the stimulus is one that is relevant to the leaf in particular, since its function is to utilize the energy from sunlight. So much do we take it for granted that plants grow towards the light that we seldom notice it outdoors. It is much more readily noticeable in house plants. If the plant is placed some little way from a window and not turned, it will be found that, quite literally, the whole plant will gradually turn its back on you and head for the light.

In some plants there is a mechanism to prevent the leaves from being scorched by too bright light; these are in the main plants that normally grow under low light intensities. Two commonplace examples are wood sorrel and the house lime. The mechanism here is for the leaves to droop under intense light in order to avoid mechanical damage to the chloroplasts.

The efficiency with which the leaves of plants arrange themselves in relation to sunlight would be of little use to the plant as a whole if the roots were not similarly sensitive in their environment.

Roots supply plants with two essential requirements (beyond the obvious one of forming a mechanical anchor). They supply the nutrients which the plants need in the form of weak solutions of mineral salts and they supply the leaves and stems with water, needed primarily to run the photosynthetic process but also to keep the cells turgid (i.e. to prevent wilting). The actual amount of water a plant needs is enormous, especially when it is realized that although many plants are over ninety per cent water, most of the water taken up by the roots is quickly lost by transpiration (the giving off of water vapour as a by-product of photosynthesis). Estimates as to just how great the throughput of water is vary enormously, but one reliable estimate suggests that in a cool temperature region over 500 tons of water are transpired by one acre of grass in the period May to July, the resultant dry-matter hay crop being a mere one-and-a-half tons. Put another way, the amount of water retained by the plant is less than one per cent of its uptake.

Roots are by nature geotropic; that is they respond positively to the pull of gravity. They grow downwards, at least to begin with. However, to survive they must have not only water but also air around them, and if they were to penetrate the soil too deeply they could reach depths at which little or no air is available for their survival. They therefore have to grow sideways in their search for moisture. The way in which a root moves through the soil is one of the marvels of vegetable engineering. It advances slowly through the soil with a gently circling motion, which enables it not only to drill its way through the soil particles but also to take evasive action if it meets an object such as a rock which it cannot penetrate; the spiralling motion enables it to find a way round. The tip is an extremely delicate and tender organ, and it is invariably protected from mechanical damage by a sort of cap or thimble of cells which are continually being discarded and replaced. The walls of these cells are mucilaginous, actually lubricating the tip of the root as it moves. The root tip has to keep perpetually advancing in its search for moisture since it is constantly drying out the area through which it has already passed. Little if any actual absorption takes place in the root tip: the area of greatest absorption is immediately behind the root tip, and can be seen under a microscope to be a dense mass of root hairs: these are constantly being produced as the tip extends, and constantly dying once their usefulness is over. These hairs actually 'taste' the soil and by increasing or diminishing the rate of growth on one or other side of the root guide the tip towards that soil which best suits the needs of the plant. As the tip pushes its way through the soil the root behind it thickens, helping to force the root tip forward.

Such complexity suggests a considerable degree of feeling in the root. In the first place the root tip is plainly able to feel the pull of gravity. Once the roots have made their initial downward growth and started to level off, it is gravity that determines the depth at which they grow. It is gravity which keeps the roots of ground elder or gout weed at so remarkably a constant depth below the surface of the soil and gravity which is responsible for the constant level at which the rhizomes of irises grow. Plainly the root hairs are able to feel the presence of water in the soil, and so avoid moving into particularly dry patches: and that the root hairs are able to taste the suitability of the soil and steer the root tip again implies a degree of feeling which few of us would suspect when we look at something so unpromising as the roots of a cabbage.

Touch seems an extraordinarily human sense, and yet it is found in most plants. It is touch, for example, that tells a root tip when it has come up against a barrier which it cannot penetrate (such as a rock) and it is touch which tells it where the soil is softest and easiest to penetrate. Perhaps the most dramatic examples of the ability of plants to respond to touch occur in the tendrils of climbing plants, and to a lesser degree in the stems of twining plants. These tendrils make wide, sweeping movements in search of a support to which to attach themselves. This circling motion, though not very obvious when one is observing the plants with the naked eye, is extremely obvious and even dramatic when revealed by a time-lapse camera. Once the tendril finds a support, it wastes no time in attaching itself. Similarly with twining plants the growing tip circles widely until it finds a suitable support, and then begins to twine quite rapidly. The mechanics of what happens to a twining stem or a tendril as a result of touch are well understood. The growth on the outer side of the tendril or twining stem increases enormously – by up to 200 times the rate of the growth of the rest of the plant while the inner surface of the coil frequently stops growing altogether. This growth differential often extends backward down the twining stem or tendril for a distance of several centimetres. With the tendrils of Passion flowers, for example, the coiling extends back from the surface round which the tendril is clinging to the stem from which it was initiated. The result of the backward coiling is to produce a flexible springlike structure which enables the stem to move in wind or under pressure from rain without tearing itself away from the coiled tendril, as it might if it were connected to it by a rigid growth.

The sensitivity to touch of tendrils and twining shoots is really quite remarkable, yet it seems that touch alone is insufficient to start the encircling growth: there must also be an element of rubbing. It is as though the plant needs to make sure that the surface it has found is sufficiently permanent for it to cling to before it commences clinging. Such rubbing occurs very naturally under the influence of air movement, and there are very few plants that will grow where the movement of air over them is less than four mph. Even wind speeds as low as this can cause sufficient movement to produce the necessary rubbing.

The reactions of many climbing plants to touch is often extremely rapid. In general tendrils react faster than twining stems, and the world speed record is held by the tropical climber *Cyclanthera pedata*, whose tendrils begin to curl within twenty seconds of their finding a support, the first complete encirclement being achieved in less than four minutes.

Responses like these show that without question various parts of plants are able to feel: they are aware of external factors through physical sensation. Of even greater importance is the way in which the individual cells seem to be able to communicate between themselves. The climbing gazanias, *Mutisia spp.*, provide a good example. These are leaf-tip tendril climbers. If a certain number of tendrils in succession fail to find a support the stem stops growing and a new shoot is initiated further down to set off in search of a support. Plainly some kind of internal communications system is at work here. If there were no such system there is no way in which the root hairs would be able to steer the root tip in the way they do. Since there seems to be a direct relationship between the complexity of structure and its ability to feel, it seems logical to assume that flowers, which are by far the most complex structures on higher plants, should be capable of both a wide range of feelings as well as considerable intensity of feeling.

If we accept that the only way in which we can judge whether plants feel or not is by the way they react to that feeling, then there is no doubt at all that flowers can feel. Unfortunately we have no way of measuring the intensity with which they feel, which is a pity, since, as flowers are the sex organs of plants, we have no way of knowing whether they derive as much satisfaction from their sex lives as we do or not.

Many phenomena appear to confirm feeling in flowers. For example many emit an odour in order to attract an insect to pollinate them, and cease emitting that odour once mating has taken place. For this change to happen the flower must be able to feel itself being mated: if it did not feel this it would not be able to react to it.

In the house lime, *Sparmannia africana*, there is a central boss of radiating stamens or male organs which immediately move rapidly outward if triggered by a visiting insect, showering it with pollen. In the barberry, *Berberis spp.*, a similar mechanism exists, except that the stamens spring inwards instead of outwards. Indeed, there is a wide variety of structures in flowers whereby insect visitors trigger off an ejaculation of pollen in a manner which closely resembles an orgasmic contraction, and these cannot be explained in merely mechanical terms. The important point is that changes take place in the flower after pollination which can only be accounted for by accepting that flowers are aware of the sexual activity in which they take part.

All this strongly suggests that various complex organic structures on a plant can feel. A question much more difficult demonstrably to answer is whether there is any centre to which or through which the messages which convey information of physical sensation pass. There is a growing body of evidence which seems to suggest that it is at least possible, if not proven, that there is, in most higher plants, some form of centralized communications network.

The sensitive plant, *Mimosa pudica*, is a case in point. These plants not only sleep, but are quite remarkably sensitive to being touched – hence their popular name. The leaf on the sensitive plant is of the form described by botanists as bipinnate. That is to say that there is a central midrib (which is an extension of the leaf-stalk) and that on either side of this midrib there are a number of leaflets, each of which in its turn is composed of a midrib with tiny leaflets along its length. If the plant is touched very lightly on one of these leaflets all the leaflets on that side of the midrib will fold up. If one segment is touched rather more vigorously, the whole leaf will fold up, while if the same segment is touched with violence, the whole plant will fold itself up, the leaflets, the leaves, the leaf-stalks and finally the stems of the plant drooping, in that sequence.

The mechanics of this operation are well understood. At the base of the leaf-stalk in nearly all higher plants there is a swollen area called the pulvinus. It is the pulvinus that is responsible for the deliberate movement of the leaves in most plants: it causes the leaves to turn towards the sun, and also initiates and terminates the period of sleeping. What happens in the sensitive plant is simply an extension of the use of this organ. The pulvinus of the sensitive plant contains cells with very thin walls, which surround the flexible lower strands in the otherwise stiff leaf stalk: these flexible strands are to a stiff stalk very much what muscles and tendons are to creatures with skeletons. They enable the leaf-stalk to be raised and lowered much as the biceps enable the forearm to be raised and lowered.

Stimulation of any part of the leaf causes water to move from the pulvinus to empty spaces, thus reducing the pressure inside the pulvinus and so allowing the flexible strands to lower the leaf. The more violent the stimulation the greater the quantity of water that moves from the pulvinus to the empty cells, and the greater the degree of collapse in the plant. A simple matter of hydraulic engineering. After the lapse of a certain period the water moves back into the pulvinus and rigidity is restored.

It is quite reasonable to assume, in the light of other plant reactions at which we have already looked, that the sensitive plant only folds its leaflets up in response to touching because it is able to feel itself being touched. While it is easy to understand how the leaflet that is being touched should react, the fact that it is possible for the whole plant to react suggests that there must be some inner communications network by means of which excitation of only one leaflet can be relayed to the rest of the plant. There is really no other reasonable way of explaining the total collapse of the plant. But this does not require one to hypothesize anything equivalent to a brain in plants.

Now if the sensitive plant can show so dramatically that excitation of one leaflet can affect the whole plant, it seems once again well within the parameters of what we have already established about feeling in plants, to assume that excitation of the sexual organs of the plant could be relayed through the rest of the plant. The style or vulva of the flower shows that it is aware of a pollinator settling on it by the way it reacts: in which case there seems to be no reason for assuming that the stigma or vagina of the flower cannot feel itself being penetrated by the pollen tube pushing its way up it.

One of the most convincing factors militating in favour of plants being aware of their own sexuality and sexual performances is what appears to be intent on their part. The commonplace hedgerow wild flower, popularly known as lords-and-ladies ensures its pollination by adopting both devices: it not only emits a scent calculated to attract the owl-midges which pollinate it, but the temperature inside the flower rises by some 15°C. In the related aroid *Arum orientale*, coming on heat is marked by an even greater temperature rise, the temperature inside the flower being as much as 30°C or more above the ambient temperature outside.

It could quite reasonably be argued that coming into heat may not in itself be a pleasurable sensation, or even something of which one is particularly aware: yet a necessary accompaniment of oestrous is heightened desire. If one were to pursue this argument one would also be led to conclude that in humans orgasm is merely the mechanical outcome of the mechanical friction of one organ rubbing within another. Few people would happily accept that the pleasure to be derived from copulation can be explained away so simply. Certainly deep sexual experience involves a great deal more than simply the sum of the parts involved, and most people would attribute the extra dimension to some inner life, such as one's emotional life. Which naturally leads one to enquire whether plants have some sort of emotional life.

Over the last couple of decades a reasonable body of experimental data has been accumulated which seems to suggest that there is some inner life to plants which is greater than anything which can be simply explained by the sum of the cells of which they are made.

Clee Backster, America's foremost expert in the use of lie detector devices, made some curious and much publicized discoveries. One evening in 1966, having nothing better to do, he attached the electrodes of one of his lie detector machines to the leaves of a *Dracaena* growing in his office: it seems he was curious to see how the plant would react to water being poured on its roots, and how rapidly. The plant responded rapidly, but instead of drawing an upward curve on the graph, indicating a greater degree of conductivity in the more turgid plant, it traced a downward curve, with a lot of saw-tooth motion.

A galvanometer is a relatively crude instrument which measures the electric potential in a human when he experiences stimuli, whether emotional, intellectual, visual or audible. It then causes a needle to move across a calibrated face, to trace the potential on a piece of graph paper. This diabolic device was invented by a Viennese priest at the end of the eighteenth century with the heavenly name of Father Maximillian Hell, SJ. When used for police work, it has been found that the most effective way to trigger the needle fluctuations that show strong emotion in humans is to threaten their well-being. Backster wondered whether he could trigger a fluctuation of the needle by threatening the well-being of his *Dracaena*. So he dipped one of its leaves in his hot cup of tea: the response was insignificant. He pondered the problem for a few minutes, and decided to burn with a match the very leaf to which the electrodes were attached. The moment he had the image of the flame and the leaf in his mind, the polygraph tracing of the plant showed a strong upward surge. Backster felt decidedly uncomfortable. Had the lie detector been reading his mind, instead of him reading its mind? He left the room to fetch matches, and on his return found that the plant had registered another surge. When he set about burning the leaf there was another, but lower peak. When he went through the motions of pretending to burn the leaf there was no reaction at all. Amazed,

Backster concluded that the plant could differentiate between his real intention to burn the leaf and his pretended intent to burn the leaf.

Backster had, in all probability, not really expected his casual attachment of electrodes to his *Dracaena* to reveal anything much, but it had, and he tried to explain it. After turning over several hypotheses, Backster concluded that 'Maybe plants see better without eyes, better than humans do with them.' He felt that the five senses of humans are a limiting factor, overlying and masking a very much more primitive universal way of perceiving. He considered that in humans the limitations imposed by the five senses are necessary. As he put it 'If everyone were in everyone else's mind all the time it would be chaos.' It would. There is a kind of peculiar corroboration of Backster's hypothesis, one which he very possibly did not know about. Totally independent research has shown that on average there are thirty per cent fewer passengers travelling on trains involved in serious crashes than the average number who could be expected on that train. That is fact. What is lacking is explanation. Perhaps there is some universal means of perception below the gate of our consciousness, and perhaps also some parallel means of communication of which we are unaware, possibly at cell or even at molecular level.

Backster's final set of experiments, like his first, came about almost by accident. He happened to notice one day that when he cracked a raw egg to feed it to his dog, a plant wired to a polygraph reacted violently. Curious, he did the same thing the next evening, with the same result. Curious to find out what the egg might be feeling, he attached it to a galvanometer. What he found was that he got recordings that corresponded to those of the heartbeat of a chicken embryo. Yet the egg was unfertilized. Later he dissected the egg, and could find no trace of any sort of circulatory system, nothing like a nervous system, nothing, in fact, that could explain the regular pulsations that so closely resembled the heartbeat of an embryo chicken. In a further experiment he attached one of two eggs to a cardiograph and dropped another into scalding water: the egg attached to the cardiograph reacted violently to the death of the other egg. What was true of plants, was true of animal life, albeit that animal life was unfertilized chickens' eggs.

Many people have followed up Backster's work, and some of their findings seem quite extraordinary. While they do not prove that plants have emotions, they at least make the question worth asking, where formerly one could have dismissed it as totally improbable.

It is all very well to know that plants have molecules similar to those found in animals, and to be able to demonstrate that they can feel by showing that they react to feelings, yet the most pertinent proof that they can feel arises only when we enquire to what purpose they feel. Since in nature nothing evolves nor is elaborated that is not essential to survival, whatever feelings a plant may experience must bestow some survival benefit. What we need to remember is that it is only man who considers that the survival of the individual matters: throughout the rest of the animal kingdom, and throughout the whole of the plant kingdom it is the survival of the species that matters: the individual is unimportant.

In the human and other higher animals an electrical unit of information can travel along a nerve at a speed of 10,000cm/sec. Reactions as rapid as this are of tremendous survival importance. Such speed of communication within the organism enables movements of quite amazing precision and coordination to be made, helpful to both the hunter and the hunted. There can be no question but that a centralized nervous system together with a rapid information return system bestows considerable survival benefits on animals.

A plant lives its life in a totally different way, at a totally different pace. While a unit of information can travel along a human nerve at 10,000 cm/sec, the rate of reaction in the sensitive plants, one of the fastest movers in the plant kingdom, is a mere 3 cm/sec.

It is doubtful whether the transmission of units of information at any greater speed would serve a useful survival purpose. Plants lack the power of locomotion, are easily damaged by browsing animals, hungry caterpillars, by wind, storm, rain, hail, snow or lightning. There is no survival benefit in a plant feeling fear at the approach of a hoard of voracious caterpillars, since it lacks the ability to flee them; nor would it serve any survival purpose for them to feel pain each time a leaf was bitten.

Plainly whatever systems of internal communications plants have evolved have arisen under entirely different circumstances from those observable in man or the other higher animals, to serve a very different purpose. Plants have no need of a high-speed internal communications system. And if they do not need that, then they do not need a central clearing-house for information as do animals.

This leads logically to the conclusion that plants respond most to those feelings which help them to survive, the positive feelings, rather than the negative ones. In the slow pace of their daily lives the leaves turn towards sunlight, the roots spiral towards softer, richer soil, tendrils reach and climb, not for the benefit of the individual but merely as a means of building the plant up till it has the strength to perform the one purpose for which it exists, to reproduce itself. Since reproduction is more important to it than its own survival, its strongest and most intense feelings are those concerned with its reproduction, primarily the pleasurable sensations of its sex life.

PART TWO
TOWARDS COITION

4
Preliminaries

A CHAPTER LIKE THIS really should not be necessary in a book like this, since it has nothing whatsoever to do with sex. It is included simply to make the going easier in the coming chapters for anyone who, like myself, is never quite sure without looking it up, whether the Oligocene period came before or after the Quaternary, and which, if either, belonged to the Mezozoic or Palaeozoic. (In fact neither was either). Which underlines the necessity for the inclusion here of some sort of ready-reference: to which end the whole 4.7 billion years since the earth's crust formed is summarized on a single page.

Similarly, while most of us know a moss when we see one, and most of us are also able to tell the difference between a moss and a seaweed, or a seaweed and a mildew, not quite so many of us are certain which of those is a bryophyte, and which a thallophyte. To clarify the confusion, the whole plant kingdom is here revealed in a diagram, startling only in its simplicity.

It should be borne in mind, however, that the division of the plant kingdom into categories in this way is purely something imposed upon the plants by man, with his passion for superimposing order on everything – except his own affairs. The categories are entirely artificial, and the plants themselves do not always seem entirely at home in their categorical pigeon-holes.

GEOLOGICAL TIME SCALES

Million years ago	*Geological periods*		*Plants*	*Animals*
2.5	Quaternary	TERTIARY		Man
7	Pliocene	TERTIARY		
26	Hiocene	TERTIARY		
38	Oligocene	TERTIARY		
54	Eocene	TERTIARY	First grasses	Early primates
65	Palaeocene	TERTIARY	First monocots	
136	Cretaceous	MESOZOIC	First dicots	
190	Jurassic	MESOZOIC		Birds
225	Triassic	MESOZOIC	First conifers	First mammals
280	Permian	PALEOZOIC		Reptiles, insects
345	Carboniferous	PALEOZOIC	First seed plants	
395	Devonian	PALEOZOIC	First ferns	Amphibians
430	Silurian	PALEOZOIC	First vascular land plants	Fishes
500	Ordovician	PALEOZOIC		Ammonites
570	Cambrian	PALEOZOIC		Trilobites
Billion years ago				
1	Pre-Cambrian		Sexual reproduction starts	Multicellular invertebrates
2				Plant-eating animals
			First blue-green algae	
3				
			First life forms	
4				
Origin of Earth's crust (4.7 billion years ago)				

THE PLANT KINGDOM VERY SIMPLIFIED

	Group *Years ago*		*Examples*
METAPHYTA (True plants with stems, roots, leaves etc)	ANGIOSPERMS (Flowering plants) 135,000,000	MONOCO- TYLEDONS	Grasses, lilies, orchids pallus, etc
		DICOTYLEDONS	Buttercups, daisies, peas, magnolias etc
	GYMNOSPERMS	CONIFERS	Firs, pines, yews, junipers, etc
		GNETALES	Welwitschia
		GINGKOES	Maidenhair Tree
		CYCADS	Cycads
		BENNETTITALES	Fossil cycads
		PTERIDOSPERMA	Seed ferns
		PTERIDOPHYTA	Ferns, horsetails and clubmosses
		BRYOPHYTES	Mosses and liverworts
PROTISTA (Plants with undifferentiated bodies)	THALLOPHYTES (Simple plants) 600,000,000	FUNGI a) Phycomycetes	Includes mildew and common bread mould
		b) Basidiomycetes	Includes rusts, sumts, toadstools, mushrooms, puff-balls & stinkhorns
		c) Ascomycetes	Includes yeasts, pink mildews & bread mould
		MYXOMYCETES	Shine moulds
		CHAROPHYTA	Stoneworts
		ALGAE	Seaweeds varying from the highly evolved kelps to primitive single-celled plants such as diatoms
MONERA (organisms without nuclei)		CYANOPHYTES	Blue-green algae
	More than 2,000,000	BACTERIA a) Photobacteria b) Chemobacteria	Bacteria

5
The Lowest Forms of Life

THE VERY LOWEST FORMS of life neither creep nor crawl. They merely exist. They are just single-celled organisms, so simple that there is no way that you can say of some of them that they are plants, and of others that they are animals. They turn up in botany text-books under one name, and in zoology text-books under another name. You can't sink much lower than that.

While it is difficult to see how anything as simple as this could possibly enjoy any sort of sex life, the fact is that the whole story of sex not only in plants but also in animals, including man, begins right here.

None of these single-celled organisms was able, nor is able, to reproduce sexually. Which does not mean that they were not able to reproduce at all. Their method was as simple as the cells themselves. They merely divided themselves into two cells, where formerly there had been one. Multiplication by division. Which is, after all, the fundamental principle of all reproduction, whether sexual or not.

It is important to understand what we mean by reproduction. Any cell that is able to produce another cell is reproducing. The result of such reproduction may not be a new individual: the new cell may merely replace an old or damaged one, or add to the bulk of the organism. It is only where the cell and the individual are synonymous, that reproduction results in a new individual.

In the lowest forms of life the cell and the individual are synonymous. So at this level all reproduction involves the creation of new individuals.

6
First Advances

SPLITTING ONESELF in half may be a perfectly adequate way of reproducing if one is a single-celled organism. The same simple method runs into considerable complications when one becomes a multi-cellular organism.

A single-celled organism consists of a single cell – a protoplast – contained within a membraneous outer casing. When it reproduces by splitting itself into two parts, protoplast and membrane are shared equally among two offspring.

When a group of single-celled organisms get together to live in a loosely organized structure, each of those single cells has to lose some of its individuality and independence for the greater good of the whole. While each cell retains its own cell walls, the whole colony is enclosed within the outer membraneous casing. When a single cell within such a colony reproduces by splitting itself in two, all it achieves is an increase in the overall bulk of the colony. It does not produce a new individual. Some other method had to evolve to achieve that.

The method that evolved is merely a modification of the principle of a single cell splitting itself in two and involves notable innovations. Somewhere on the periphery of the loosely organized colony, the protoplast of one cell begins to multiply, and in so doing ruptures the cell walls, and floats out into the surrounding medium, where it continues its rapid series of divisions, these eventually leading to the creation of a new individual.

At this stage in evolution there is no way you can tell in advance a cell which is going to reproduce from one which is not. All the living cells in the individual are the same size, the same weight, and, to all appearances identical in all respects. In principle any of the cells would contain sufficient genetic information to reproduce the whole plant. There seems to be total equality.

A cell which functions as a multiplier cell is called a spore. It is the simplest type of spore there is, merely a protoplast that has escaped from its cell walls. Its evolution opens up whole new worlds, vast vistas of evolutionary possibilities, including ultimately the joys of sex, which would not have been possible had simple body fragmentation been the only way forwards.

7
Sex in Season

SPORES ARE NOT, even among primitive plants, produced continuously like sperms in man. They are produced seasonally in response to specific conditions.

Among the single-celled algae it is not uncommon to find that, when conditions become unfavourable for vegetative growth, the cell wall thickens. When conditions once again become congenial for growth, the thickened wall is cast, and a new wall is grown. Among bacteria, another very primitive plant, the reaction to unfavourable growing conditions is different and more intriguing. They completely shed their membraneous walls, and the naked protoplast rounds off, condenses and goes into a dormant stage. When conditions favourable to growth return, it expands, grows a new membraneous wall and continues to live as before. In both cases the purpose of what happens is protective, not reproductive, yet the processes are in many ways analogous.

Essentially the same thing happens with spores, except that when conditions become propitious for growth, the spore sets about building a whole new plant. Plainly, since most cells in primitive plants can function either as vegetative or reproductive cells, there is something about adverse conditions which inhibits continued vegetative activity. Expressed more strongly, the production of spores is a response to a threat to the well-being of the parent plant.

The next logical step was the evolution of specialized spore-producing cells. The process was gradual. The earliest such cells functioned as normal vegetative cells for most of their life, only taking on their specialized spore producing role once conditions for vegetative growth declined. Only eons later did cells evolve whose sole purpose was to produce spores. Indeed, within the living algae, it is possible to arrange a continuous series from those in which all cells can function either vegetatively or as spore-producing cells, to those in which the majority of cells can function only vegetatively, while a relatively small number have become specialized spore-producing cells called sporangia.

In a sex-obsessed world it is not altogether without interest to note that the very first specialized cells to emerge in the plant kingdom were the reproductive cells.

The arrival on the evolutionary scene of specialized reproductive cells led to a subtle change in the whole body of the plant. Those cells concerned with reproduction became more specialized, while those not so concerned became simpler.

At this stage, some two billion years ago, evolution had reached the point at which there were simple algae that produced simple spores that in turn produced simple algae and so *ad infinitum.*

The next step would appear also to have arisen in response to conditions unfavourable to normal growth. What happened was that the sporangia, instead of producing every spore the same size began to produce spores of different sizes. The spores produced closest to the period of optimum growth had more time and more suitable conditions in which to multiply than those produced nearest the end of the period of active growth, and so were larger than those produced later. Which may not seem of much interest but it was in fact one of the most momentous events in the whole star-studded panorama of evolution. Though it may not yet be quite clear why.

8
The Sexual Awakening

SEX is not something which burst all of a sudden upon an unsuspecting world, ushered in with a fanfare of trumpets, and choirs of pagan angels singing its praises. Nor was it invented once and once only, at one time in one place by one group of plants. Indeed, it seems rather that several groups of plants were groping in the dark, not quite sure of what it was that they were trying to find. It grew gradually out of the differentiation of spores into large ones, small ones and in-between ones.

The situation was not, however, entirely satisfactory. The large spores, rich in nutrition, germinated readily enough, but the smaller spores found it increasingly difficult and the smallest spores of all simply lacked the nutritive capacity to germinate at all.

After some eons of this highly unsatisfactory state of affairs, some of these tiny little impotent spores discovered that if they paired up with a chum, then the two of them together could fuse into a single spore which did have the power to give rise to a new plant. Once this discovery had been made, it became quite the thing for impotent little spores to pair up together.

It is this fusion of two cells, each of which on its own lacks the power to give rise to a new individual, but when fused together can do so, that is the most primitive sex act in the world.

Which leads to the startling conclusion that the first sex act that ever took place on the face of the earth was the result of a homosexual relationship between two impotent spores.

It is a sobering thought that had not these mindless morons of the vegetable kingdom discovered sex no further evolution would have been possible, and we ourselves should not be here to cogitate upon these mysteries. There is a little more to this than meets the untrained eye. A spore, by definition, cannot fuse with another spore. A spore that can fuse with another spore is no longer a spore: it has evolved into something else and that something else is called a gamete.

Gametes evolved directly from spores. They are structurally

identical. One very evident clue used in diagnosing the relationship of spores and gametes are the cilia, or swimming tails. If the spore has four, then you can be sure that the gamete also will have four; if the spore has fifty, then so will the gamete; if it is a poor, deprived and under-privileged spore with only one swimming appendage, then the gamete too, will be poor, deprived and under-privileged with only one swimming appendage. This occurs in genus after genus, species after species.

Curiously, when sex was first invented plants did not regard it as a particularly reliable means of reproduction, since they maintained the ability to reproduce by spores as well for a considerable period. They hedged their bets, and perhaps wisely, for some of the early experiments in sex were somewhat abortive.

It might seem from this that the primitive plants among which sexual cells were first invented, produced both spores and gametes at the same time. This was not the case. A typical life-cycle shows that during phases when conditions are ideal for vegetative activity, all the cells in the plant put all they had into growing. It was only when conditions became less propitious for vegetative activity that some of the cells started producing spores; and it was only when conditions became inimical to the production of spores, that the spavined little gametes were produced – the last fling of the plant terminating its activities for the season. Of course there must have been some overlap from one phase into the next: the plants did not perform by numbers like little soldiers learning to present arms. The production of spores phased in gradually as the vegetative activity waned; there then ensued a period of spore production before the gamete production was phased in. Over millenia the smallest spores no longer had the capacity to germinate at all. It was at that point that they resorted to the unseemly procedure of pairing and fusing. There is no precedent for such outrageous behaviour in the plant kingdom.

Gametes like true spores, have cilia or tails – very much like those which propel a sperm towards its destination. However, while spores use their cilia to move themselves away from the parent plant, gametes use their cilia to bring themselves into contact.

There is nothing haphazard about the coming together of two gametes. They do not simply collide with one another like people in the rush-hour. Using their cilia, both to propel and steer themselves, they approach each other until they come into contact; their positions with regard to one another are precise, the ciliated ends touching. And then they fuse as easily and completely as two drops of mercury, with the precision of space ships docking.

We know how they come together and fuse. What we do not know is why they come together and pair. It is not as though there were any attraction between male gametes and female gametes. At this stage of evolution gametes are gametes, neither positively male nor positively female. If anything, they tend to err on the side of being male, the maleness or femaleness of cells in the plant world being related to nutritive capacity, the capacity of these primitive gametes being very little, as is evidenced by their lacking sufficient nutritive power to germinate as ordinary spores.

To us it seems only natural that male should pair with female; at least, it is usually only such heterosexual relationships that produce offspring. It would be a very queer relationship in which the mating of male with male could give rise to offspring. In fact, there is something unexpected about the progeny of such a coupling. When two gametes fuse they do not burst immediately into growth, but raise their defences against unfavourable conditions. The fused gametes form a single cell called a zygote. This remains dormant until conditions are once again favourable for vegetative growth. Then the zygote gives birth to a swarm of spores, it being these spores that in their turn produce the new individuals. Thus the original purpose of sex was protective, not reproductive.

Which may make the whole sexual exercise seem rather pointless until one realizes that what has been achieved is the successful safe-keeping of genetic material through a period of adverse conditions. Other benefits will accrue later, but this was the first, the most important, and ultimately, the most lasting.

In the earliest of the sexual phases, gametes are neither male nor female. They are only sexual in their ability to fuse. Their differentiation into male and female followed exactly the same pattern as the evolution of sex itself. Sex itself arose because some of the spores lacked sufficient nutritive capacity to give rise to new individuals on their own, and could only do so by fusing with other similarly spavined little spores. The others germinated without pairing.

Over a pretty extended period of evolution some gametes gradually increased in size, while the others did not. Those that became larger eventually increased their bulk to many times that of the other gametes. Since there is no way in which the nucleus can increase in size, what increased in size could only be the cytoplasm, which is the nutritive tissue surrounding the nucleus. In the vegetable kingdom increased nutritive capacity is a singularly female attribute, and it was those gametes which increased their nutritive capacity which became the female gametes or, to bring them back within the sphere of normal language – eggs. However, as is so often the case in evolution, an organism gains one survival characteristic at the expense of another. In this case the price that the egg paid for its greater bulk was the loss of its cilia or swimming tails. A passive female is a much easier object to locate and penetrate than would have been a moving target.

The male gamete did not turn into a sperm simply by default. It, too, underwent evolutionary changes, but of a reverse nature. It increased its swimming efficiency at the expense of its cytoplasmic bulk. The two processes were complementary: the decrease in bulk would have given the sperm (for that is what it has now become) a greater power/weight ratio, even if there had been no increase in the efficiency of the swimming appendages. The price the sperm pays for its greater mobility is a reduction in its life-span – this being directly related to the amount of nutritive cytoplasm present.

Nature did not rest, having evolved both sperms and eggs, but continued to elaborate the theme, playing endless variations on relatively limited materials. The egg remained for a long time relatively constant in its form. But the elaborations that occurred among the sperms are quite amazing.

9
Disposable Sex Organs

THE EARLIEST SEX ORGANS ever were merely transient appendages on the surface of the body proper. There is a slight problem in explaining the antecedents of these organs, since there was nothing like them before them from which they could have evolved.

The original function of sex organs was protective rather than reproductive, as it was with the sperms and eggs themselves. They arose in response to the declining vegetative activity of the plant. The temporary, indeed disposable, nature of the sex organs seems to confirm this. Really there is much to be said in favour of disposable sex organs that can simply be discarded once they have served their function: they are certainly less of a liability than the permanent organs upon which man so greatly prides himself.

The extraordinary thing about these sex organs is their quite remarkable superficial resemblance to the sex organs of the higher primates. The male reproductive organ (or to be more precise, the organ producing the sperms) is quite remarkably phallic in shape, while the female organ is spherical, much as is the womb of woman. The female organ is large, well-fed and well-developed, while the male one is rather small and insignificant. The huge female organ contains a single egg, while the trivial little male organ produces an enormous quantity of sperms.

At which point, after some 5 billion years of evolution, we have reached plants with which most of us are more or less familiar. The algae include among their vast numbers the sea-weeds. There are two which are familiar denizens of the beaches in most temperate areas of the world, the wracks and the kelps, and they illustrate very well both the most primitive method of sexual reproduction among algae, and also a slightly more advanced method.

The wracks are a familar sight along the coasts, especially where there are rocks, for they cling to rocks, by means of a very primitive cement which they secrete, between the levels of high tide and low tide. The sex organs of these plants are among the

most throw-away type known in the plant world. Both the male and the female sex organs consist of what may loosely be described as an envelope, one containing female cells, the other male cells. The envelopes themselves are contained in pouches produced on the body of the wracks. When the tide goes down, leaving the wracks high and dry, the wracks shrink in the dry air, and this shrinking forces the envelopes out of their pouches. Once they have been forced out in this way, they remain attached to their parent wracks by a water-soluble glue (orange for the male, green for the female – it seems that colour-coding is nothing new). When the tide comes back the sheer force of the water bursts open the envelopes, releasing both eggs and sperms into the water, at the same time dissolving the glue which held the sex organs attached to the wracks. These disposable organs then perish at sea. Meanwhile the sperms and the eggs are being swirled around by the tides and currents, which must get them thoroughly mixed-up. The eggs are, of course, completely passive, having abandoned their swimming tails in favour of greater nutritive capacity. All they can do is to let themselves go with the tide, frantically emitting a kind of chemical 'scent', designed to attract the sperms. Meanwhile the sperms, which have gained in mobility by elaborations of their swimming tails at the expense of their bulk, are lashing around in the waters trying to locate the source of the chemical 'scent'. While the eggs can float around passively for a relatively long period, things are slightly more desperate for the sperm, since its nutritive capacity is strictly limited. In effect, its only hope of survival is to mate with an egg before its nutritive capacity has been exhausted.

Thus, propelled by their cilia or flagella, the sperms hunt the egg. Each egg is capable of attracting many sperms. These attach themselves to the outer wall of the egg, and continue thrashing the water with their tails in their efforts to push themselves through the wall and into the egg. As a result of all this thrashing of flagella, the egg begins to spin, and spins faster and faster the more sperms attach themselves to it, though whether this spinning is an integral part of the sex lives of these wracks or purely incidental, is not known. Eventually some lucky sperm penetrates the outer wall of the egg, and the two nuclei fuse. Since the essence of sexual reproduction is the fusion of one nuclei with another, only one sperm can penetrate the egg. Which leaves all the other sperms thrashing their tails to no effect. To prevent them exhausting themselves for nothing, the chemical 'scent' of the egg changes once the two nuclei have fused, this changed scent telling the other sperms that they have lost that round. The sperms then quickly unhitch themselves and go off in search of some other egg.

The fertilized egg then divides itself into two cells, then four, then eight and so on, each of these new cells being exactly like the original fertilized cell, which in itself was exactly like any other cells from the parent wrack. And all of its life all that this new plant will do is go on increasing its bulk by cell division until it, too, reaches sufficient size to be able to produce sex organs. And so the cycle goes on.

Gradual modifications were made to this system over the next few eons in the kelps, another familiar seaweed of temperate beaches. In the kelp, one has what is probably the most sexually sophisticated of all the seaweeds – something one might quite easily overlook as one treads a lowly kelp underfoot.

Among the kelps the pouches containing the sperms and eggs occur on the tips of the plant, at the outermost limit of its growth. These, when fully developed, are washed loose from the kelp. Once separated from the parental kelp, these little things start growing, while drifting with any passing current. In growing they do not become like the parent kelp. Instead they grow into thread-like objects that float on the surface of the seas. It is these stray threads which are in fact the sex organs of the kelps. Some of these threads will produce sperms, others will produce eggs, after which the threads perish. The manner in which sperms hunt down the eggs is much the same as with the wracks.

To humans, who are quite possessive about keeping their sex organs firmly attached to their bodies, and who moreover are used to seeing most of their friends doing the same, this idea of sending one's sex organs off on a life of their own, may seem rather peculiar. Yet it bestows a survival benefit, and it is not too hard to see what that benefit was. A kelp, firmly anchored to the sea bed, is totally at the mercy of the elements: it takes a perpetual bashing from the tides, and is tugged at and pulled at by tides and currents. Disposable sex organs have a far greater chance of survival, being able to float freely on the surface of water, than would sex organs permanently attached to the plant.

Such an arrangement is one of the most primitive examples of what is known as 'the alternation of generations'. This amounts quite simply to this. You have a plant. That is one generation. And you have its detached or detachable sex organs, and that is the other generation. And the two generations alternate endlessly. That some survival value accrued from this arrangement is apparent because of its persistence. In all higher plants, even the most highly evolved of the flowering plants, this alternation of generations continues to take place, albeit in a more and more vestigial form. It is also the main road along which evolution was to travel, and it was this that made the evolution of the lowest forms of landlife possible.

However, since most evolutionary steps involve a response to some sort of problem, whether external or internal, it is worth looking briefly at the internal problems to which the alternation of generations is the logical response.

10
Sex by Numbers

THERE IS NOTHING NEW in this world about sex problems. They are as old as sex itself. If we know less about the sex problems of primitive plants, that is probably because there was no media in those far-away days to bring the anguish of an adolescent algae to our attention. The problems of those days were not the all-too familiar problems of promiscuity, moral perplexity, premature ejaculation or even of the euphemistically-called social diseases. They were far more fundamental than that. They belonged to the realms of what might be called the genetics of sex. The genetic information is a coded message carried in units called chromosomes. For every plant and animal specie, the chromosome number is fixed. It may be 32 for an apple, and 16 for a fig. It isn't but it might have been. For a kelp it is *x*. Now every sperm produced by a kelp also has *x* chromosomes. But so too does every egg produced by a kelp. So if a 1*x* kelp sperm fuses with a 1*x* kelp egg the result is going to be a cell with 2*x* chromosomes. If that 2*x* fertilised egg were to produce more sperms, they too, would be 2*x* sperms, and the eggs would be 2*x* eggs as well. So that when they fused the resultant fertilized cell would be 4*x*. The next generation would be 8*x*: then 16*x*, 32*x*, 64*x*, 128*x*, 256*x*, 512*x*, 1024*x*, 2048*x*, 4096*x*, 8192*x* and so on *ad infinitum*, with the result that the genetic message required to reproduce a no longer so simple algae would be some 357694088973645295736285967384996735463442613079584 40*x*. Such an enormous quantity of genetic information would leave little room in the cells for anything else.

There is one sense in which this is a problem produced by sex. In another sense, however, it is a problem that could only be solved by sex. The way it was solved was this. You start with an ordinary kelp – which is 1*x*. It gives rise to both sperms, which are 1x and to eggs which are also 1x. When a sperm fuses with an egg the resultant organism is 2x. And that 2x organism grows into an adult kelp as we know it. And that produces floating threads which are 1x, producing 1x eggs and 1x sperms. So that in the kelp and similar primitive plants the chromosomes are halved and doubled every alternating generation.

It was 'the alternation of generations' that made genetic inheritance and variability possible and in so doing played a crucial role in enabling plants to move out of the seas and onto what one might rudely describe as dry land.

11
Some Sexual Deviants

THE MAIN STREAM of the evolution of sex in plants is plain enough. It lay through the higher Thallophytes, that is the most advanced of the algae and the fungi, with their first tentative steps towards the alternation of generations, and on into the Bryophytes, in which the alternation of generations becomes more firmly established. There were, however, many curious deviations from the norm before this line was clearly established.

Sex, after all, is one of the most exciting discoveries most people ever make, and they do tend to go overboard on it a bit when it is new to them. So, too, did the primitive plants. Indeed, they seem to have tried just about every sexual experiment they could with their limited equipment. There is nothing at all surprising in that. The only thing that is surprising is that some of these experiments, though evolutionary sideroads to nowhere, have managed to survive into the present day.

One very logical experiment was tried by one group among the red algae. In this the fertilised eggs give rise to a body as elaborate as the parent body itself. So similar are the two phases in this alternation of generations that the two can only be told apart by a chromosome count.

Slightly higher up the evolutionary ladder, and pointing roughly in the right direction, are algae in which not merely two quite distinct alternations are found, but three quite distinct phases of body-work. Perhaps the oddest thing in this case is that there is, in a sense, no parent organism. There is a male parent, which produces sperms, and a female body, of quite different appearance, that produces eggs. An almost human arrangement. When the sperms and eggs join up, they produce a third and distinct individual which produces spores, the spores giving rise to male and female gametophytes once again.

The most curious sexual experiments were not carried out by the algae, but by the other huge group within the Thallophytes – the fungi. This is a group of plants so large that it is divided into three sub-groups – the Phycomycetes, the Ascomycetes and the Basidiomycetes.

It is within the smallest of these three groups, the Phycomycetes, or algal fungi, that a possible step forward in evolution might have occurred, but somehow did not quite make it. In them the male and female sex organs are well developed and readily recognizable. They are almost identical in structure to those of many of the algae, and seem to offer little or no significant advance over them. Nothing particularly interesting about that.

What is interesting about them is their life style, and the way this affects their sex life. Any creature that has a peculiar life style is liable to find itself leading a peculiar sex life. The particular way in which the life style of these fungi differs from the life style of most plants is that they have what has been called 'the dependent habit'. They do not sustain themselves by the normal photosynthetic methods but derive their sustenance either by living as parasites upon living hosts, or as saprophytes, drawing their nourishment from hosts who were formerly living. Their vegetative activity is therefore extraordinarily intimately linked with the life of their host. Once you have, almost literally, bled your host dry, your own life must terminate. Your only hope is the continuation of your kind by reproducing as rapidly and prolifically as you can. And this is exactly what happens in this group of fungi. When the supply of nourishment from the host begins to fail, vegetative activity necessarily begins to decline. It is the onset of this decline that appears to initiate the formation of the sex organs, and the subsequent fertilization of the egg. Once the eggs have been fertilized the plant is assured of the continuance of its kind.

Because these plants lead a peculiar life, their sex organs have had to adopt peculiar methods of ensuring the fertilization. The male organ, the antheridia, and the female organ, the oogonium, while virtually identical with those of the higher algae, differ in that male and female organs are produced in close proximity to one another, in some cases actually touching. In algae the free-swimming sperms are attracted to the free-floating eggs by a chemical scent: a similar attraction exists in these fungi. As a result of this attraction the male organ bends towards the female organ. It then grows a phallic cylindrical or tubular extension which penetrates the female organ, discharging its male nucleus into the nucleus of the female cell. The growth of this male organ, and the way it penetrates the female organ is almost copulatory it is so explicit. One of the most well-known of the fungi of this group which conducts its sex life in this way is the downy mildew.

It is the two other much larger groups of fungi, still mainly parasitic or saprophytic, that show much more typically the consequences of a dependent mode of life upon their sex life. In common with the downy mildew and its relatives, the sexual nature of the plant only becomes activated once the supply of nutrients begins to run out. Where these two groups differ is that, instead of having well-formed and recognizable sex organs, the sex organs seem to have disappeared. In the larger of these two groups the, Ascomycetes, the whole range of sex organ development is present, from cells that are virtually indistinguishable from normal vegetative cells to cells that are highly specialized and structured. It would be easy to make the mistake of thinking about what one is looking at as an evolutionary development: in fact it is a complete reversal: it is devolution, in its proper sense, not in its political nonsense. That this reduction of previously evolved organs is intimately related to the dependent habit is fairly clear. It is also important to understand the interrelationship of these two things, since reduction is a feature of flowering plants, which are themselves highly dependent.

In the second and larger of these two groups of fungi, the Basidiomycetes (the group to which the well-known rusts and mushrooms belong) sex organs appear to have been eliminated completely. In a portion of their life histories two nuclei consort together, generation after generation, without fusing. Finally they do fuse, and this fusion is followed by single-celled generations. The next obvious step in devolution would be the complete elimination of sex.

It is amongst that anomalous group, the stoneworts or Charales, that the most extraordinary sexual aberrations occur. Some botanists include them along with the algae and fungi in the Thallophytes. Other botanists include them, along with the liverworts and mosses, among the Bryophytes. Each group would admit that wherever you place them, these peculiar plants, remain anomalous.

It is the peculiar sex organs of the Charales that lead to this problem. Although they are plainly very primitive plants, the Charales have sex organs of considerable complexity, certainly far more advanced than those of the lowest forms of land life. They have no known close relatives: they lack either antecedents or lineal descendants, but seem rather to have been an evolutionary experiment that was aborted before completion. They stand on a sideline, shunted into a marshalling yard and then forgotten.

In spite of this, the group is worth singling out for special mention because (if for no other reason) the group has the most complex male sex organ in the whole of the plant kingdom. Man, in *sensu stricta* may have the largest organ in relation to his body size, but the Charales have the most complex. What is so odd about the group is that while the sex organs have developed in ways that are very likely associated with those adaptions that would be standard procedure in primitive land plants, they are in fact water-dwelling plants. The egg-producing cell, the oogonium is in general structure, basically similar to that of most algae. What is different about it is that it is invariably associated with a second cell, always placed immediately beneath it, whose function it is to produce a number of filaments which then twine themselves round and round the oogonium in a tight spiral. Later these filaments harden to form a hard casing. The purpose of such a layer of hardened tissues must have been protective. It must have evolved in response to conditions unfavourable for vegetative activity, though whether those conditions were exposure to air or to some other unfavourable condition such as seasonally brackish water, it is now impossible to assess.

The male organ of the Chara is absolutely unique in the plant kingdom. There is, literally, nothing else quite like it. It consists of a curious spherical case of heavy-walled and interlocking cells in the centre of which, arising from the encasing walls, is an elaborate succession of cells that finally ends in numerous delicate filaments. This fabulous organ, therefore, appears as a ball-like case packed with a tangle of threads. Each of these threads, of which there may be several hundred, consists of 200 cells, and each cell produces a sperm, so that the total sperm production of this fabulous male organ can be as many as 50,000 sperms. That may seem a very small number when compared with 5,000,000 produced by the average human male in an average orgasm at the peak of his performance, but it is a phenomenal number for an organ of microscopic dimensions. Further, the sperms themselves are of a highly specialized character, having a spirally coiled body, with an elongated cytoplasmic beak terminating in two long tails or cilia.

In complete contrast to the complexity of the male organ of the Charales, the male organ of the marine red algae is simplicity itself, while the female organs are relatively complex in their own way. The complexity does not involve the egg-producing cell, but a number of secondary cells associated with its function. The egg-producing cell does not in fact produce an egg: that it is an egg is apparent only in its function, not its form. The number of associated cells varies, but there is always one other that is present and necessarily so: this is the one that catches a passing sperm and provides for it a passageway to the eggless egg-cell. This wicket-keeper cell frequently looks like a hairlike extention of the eggless egg-cell.

The male organ, by contrast, is starkly simple. It is a simple, single-celled structure with no associated cells. Strictly it does not even produce sperms, since the cells it discharges are not motile. In some of the red algae this sperm-that-is-not-a-sperm is not even discharged from the male organ: the male organ simply breaks off from the parent plant and floats into the entrance of the female organ.

The sex organs of the Charales and the red algae have been singled out for special treatment because they are, in the currently acceptable terminology, special cases. They developed peculiar sex organs under peculiar circumstances. They do not represent the mainstream of evolutionary development any more than transvestitism represents the mainstream of human sexual development.

12
Problems of Exposure

THE MOVE from an aquatic environment to dry land was one of the most momentous and difficult steps ever undertaken on the face of the earth, involving complexities which we seldom have need to consider. Probably different organisms attempted it and failed, again and again, before the step was successfully accomplished. Incredible though it may seem, plants and animals, some of them quite large, lived in the oceans for some 270,000,000 years, and nothing at all inhabited the dry land.

But then, land then was a far more inhospitable environment than it is today. Indeed, it was positively hostile to any form of life. It was totally bleak, bare and barren, utterly desolate. The continents were just rock and scree, and the sun burned down on them, with never a blade of grass, not so much as the wing of a bird to cast the least shadow. It was a world of extremes: of blazing sun or drenching rains, of dramatic electric storms and almost continuous volcanic and earthquake activity. It was into a world as inhospitable as that that the first plants had to move once they left the oceans.

For a plant, adopting a land-based life style, meant very much more than simply dragging oneself up the beach and putting down roots. For one thing, roots had not yet been invented.

If you live in the water, not having roots is not much of a problem. There is water all round you, and if you need some water, you simply absorb it. Any organism that is accustomed to that mode of life would simply die of dessication on the land.

The Bryophytes – the mosses and liverworts – beat the immediate problem of dessication by developing a thin waxy coating or cuticle, covering their bodies. This prevented excessive loss of water through transpiration. The cuticle did not prevent water loss altogether, which means that it was also necessary for land plants to evolve some means of absorbing water. This the Bryophytes did. The most elementary roots were not true roots, but crude structures called rhizoids, which are little more than

thread-like anchors which also happen to be able to take up sufficient moisture from the ground to keep the plants turgid. (Plants, unlike people, need to be turgid most of the time).

While the Bryophytes adapted remarkably well to land in most respects, they made one incredibly stupid mistake. They omitted to adapt their sex organs. So the stupid things just sit there, stuck to the ground with their sex organs fully exposed to the dessicating influence of sun and wind.

Sexually, some standardization and streamlining has taken place especially in the body of the sperm. This is characteristically among the Bryophytes made up of a body, containing a nucleus (which is slightly curved), a more or less apparent but not pronounced beak, and two cilia or swimming tails, simple in design, but highly effective organs of propulsion. In fact, such sperms differ very little from fairly simple two-tailed spores. The spores of the mosses and liverworts, on the other hand, have undergone some quite radical changes. They have abandoned their swimming appendages completely, and gained in bulk by way of compensation. Compared with the sperms, the spores are huge objects – yet tiny by human standards – mere dust-like particles. While the sperms retain their swimming ability, the spores have abandoned it in favour of a novel way of getting about – that of allowing themselves to be transported by the wind.

Now the sex organs of these plants were developed on the upper surface, and were therefore just as much in danger of dessication as the body of the plant itself. They therefore made a similar adaptation, the outer cells forming a kind of jacket round the sex organs.

While a watery medium favours the gametophyte, the sexual generation, so that fertilization of the egg is accomplished by a swimming sperm, a land habitat favours the sporophyte, the 2x-generation, which produces spores, which are readily disseminated through the air. The generation which produces the spores, is thus an aerial generation, raising some at least of its body tissues off the ground into the air: the spores once disseminated, will only come to life to produce the sexual generation where conditions are sufficiently moist for the sperms to swim to the eggs. The need for the spores to be produced in dry air, and the sperms and eggs to be produced under moist conditions, would have led to even greater and greater divergence in the body form of each two alternating generations. Such a process would have tended to make them, however similar to each other they may have been in the first place, more and more unlike each other as time passed.

The sex organs of the liverworts bear a peculiar resemblance to the line drawings of the internal workings of the human female, and to a lesser extent to the external workings of the human male. Those of the mosses are similar, differing only in details. The female reproductive organ, the archegonium, is quite extraordinarily like a womb – a tiny, flask-shaped object with a long narrow neck, which not only looks remarkably like a vagina, but also serves a similar function. At the bottom of the flask the crucial egg is concealed. The male organ, the antheridium, is an erect phallic mass of cells encased in a protective foreskin of sterile cells. At the bottom are hollow sacks, much resembling a scrotum, each containing thousands of sperms. The main difference between the scrotum of the human male and that of the liverwort is that the 'scrotum' of the liverwort bursts open when the sperms are ready to be liberated, which fortunately is seldom the case with the human male. The sperms are well-equipped with tails to enable them to swim. But of course they can only swim when there is water present for them to swim in. It need only be the dampening of dew – that is all they need. Now, at the same time that the sperms are released from their sacks, the neck of the female organ opens, emitting a strong chemical scent. The sperms, once they are released, swim as rapidly as possible towards the source of this chemical scent. Assuming they manage to find it, they then swim down through the neck of the flask to the egg in the womb of the flask, which one and only one will penetrate and fuse with. The others perish. Once the egg has been fertilized the whole female organ starts to grow. The neck elongates, and the whole organ slowly turns into a long, thin, stalk capped with a little hairy, fairy-like umbrella-like hood called the calyptra. When the fertilized egg has matured, the hood falls off and the egg is blown away on the wind. It is, in fact, merely a spore. When it germinates it forms a mass of branching filaments, very much as in the kelps. New plants spring from buds in these filaments. The liverworts, as one sees them clinging to rock faces and so on, are the sexual generation: the filaments are the non-sexual generation. The sexual generation therefore starts life as a parasite on the non-sexual generation.

If the term parasite offends it is worth remembering that the relationship between the sporophyte and its host gametophyte is exactly the same as that between the foetus in a mother's womb and the mother herself: the unborn infant is parasitic on its mother from conception almost till birth, though we seldom have cause to think of it in that way.

Where the bryophytes made their great mistake was in producing their sexual organs on the top of their bodies. It is only when one carefully considers just what it is that the sex organs have to achieve, that one begins to realize what a ghastly mistake the bryophytes were making.

The purpose of any sex organ is quite simply to enable a sperm to reach an egg and fertilize it. It is this for which the arrangement of the sex organs of the bryophytes are so singularly ill-fitted. They are, however, superbly placed for the next function of the sexual union, the dissemination of the spores. Since the spores are disseminated into the air, the higher up on the plant the organ which disperses them is placed the better. This position, in which both male and female sex organs are placed on the top of the plant, favours the spreading of dry spores through dry air: but it militates viciously against the chances of the sperm, which needs a liquid medium in which to swim in order to fertilize the egg which becomes the spore. The sperms, in order to reach the egg which they must fertilize have to swim up into the highest and driest part of the plant.

Plainly those conditions which are favourable to the sperm, are unfavourable to the spore. And those which favour the spore do not favour the sperm. These demands are completely contradictory. The manufacture of food is essential for both sperm and spore. What the plant kingdom faced at this stage was a crisis.

It is at this singularly critical point that the largest lacuna in the whole history of the evolution of sex in plants occurs. The record of living plants shows the advance of the sporophyte to a semi-parasitic, semi-independent existence. In the living ferns that independence is complete. There must have been intermediate stages, but they have not survived into our world, perhaps quite simply because they could not compete with the efficiency of the system to which they gave rise.

13
Of Maidenhairs and Other Matters

THE ALLEGED 'failure of the Bryophytes', is due almost entirely to their ridiculous habit of carrying their sex organs on their backs. There is simply no other part of the anatomy of any plant that is going to lie more or less flat upon the ground that could be considered to be more ill-considered. In such a position, the sex organs are exposed to every passing or permanent problem imaginable. There is the perpetual danger of dessication, the danger of dehydration of the sperms before they reach the female organ, sun-scorch, wind-damage, especially from drying-out, and a multitude of other hazards. It is, perhaps, scarcely to be wondered that plants which place their vital parts in such an exposed place should have failed to leave any direct heirs. No doubt the Bryophytes gave rise to many groups of plants of which there are no living or fossil signs today, in their attempts to find somewhere better to keep their sex organs. Plainly, whatever group succeeded best in overcoming the reproductive problems faced by the Bryophytes, would be their inheritors as the dominant vegetation in this world – at least for a while.

Their inheritors were in fact the Pteridophytes and, the ferns, their friends the clubmosses and horsetails. They endured, supreme upon the face of the earth, right through the Devonian age, the dominant vegetation for a period of some 100,000,000 years. Their immediate precursors, plants like *Rhynia*, emerged into the primæval slime of the Silurian period, while their evolutionary heirs, the seed ferns, gradually took over from them during the Carboniferous period as the dominant vegetation, yet there are still ferns living today, usually relatively small plants – relative, that is, to the stature they attained in their heyday.

The success of the ferns was largely due to the way in which they successfully managed to rearrange the positioning of their sex organs in such a way that they were no longer exposed to all the hazards that faced the sex organs of the Bryophytes.

Perhaps from a biological rather than a sexual point of view, the

important thing is that the Pteridophytes, the clubmosses, horsetails and ferns (in that evolutionary order), are vascular plants. Vascular plants are plants in which there is a definite structure through which water and foods in weak solution can be conducted through the plant to the leaves, and the products of photosynthesis conducted back again to the roots and to every other part of the plant. The mosses and liverworts could not do this. The difference in the movement of water and nutrients from one cell to another in vascular and non-vascular plants has been likened to the movement of water slowly seeping through a swamp (in the case of the non-vascular plants) and water moving through clearly defined drainage channels (in the case of the vascular plants). This is a fundamental difference. It is not something that could have happened overnight – even in the slow, long nights of the evolutionary past. It must have taken eons to occur. For without the vascular bundles, which give the plants the rigidity they could never have got up off the ground and grown into anything bigger.

It seems almost certain, in the light of modern palaeobotanical findings, that the mosses and liverworts were an evolutionary dead end, and that the ferns evolved from a more primitive ancestor, very possibly one they shared in common with the mosses.

The great achievement of the ferns and the other Pteridophytes is that they kept their sperms and their spores quite separate, producing each in alternate generations that were ideal for the purpose. In so doing they fixed the alternation of generations as a mode of life. All the plants that evolved from them incorporated the alternation of generations into their existence: it is still there, even in the most recently evolved plants, though in very vestigial form.

What one has is one generation which produces spores and is therefore known as the spore plant or sporophyte, and one generation which produces gametes and is therefore the sexual generation, known as the gametophyte.

In the ferns, both the sexual gametophyte and the non-sexual sporophyte lead pretty much independent lives. Each is able to manufacture its own food, and each is able to carry out its own function in surroundings congenial to that function. The sexual generation enjoys the moisture it needs, while the non-sexual generation enjoys the air it needs. However, in order to achieve the most favourable conditions for each of the alternating generations, a number of changes highly significant for the future

of the plant kingdom have had to take place. The most immediately noticeable and obvious difference is that the sexual generation, the gametophyte, has become a tiny inconspicuous body, only likely to be found by people who know what they are looking for and where to look for it, while the non-sexual, sporophyte generation has become quite extraordinarily large. Indeed, the magnitude of the change that has taken place can be gauged when one realizes that when one looks at a moss plant, it is the gametophyte or sexual generation that one is looking at, but when one looks at a fern it is the sporophyte or non-sexual generation that one is looking at. Further, while the narrowly-stalked sporophyte of the mosses is wholly parasitic on the sexual generation, in the ferns the sporophyte is parasitic on the gametophyte only for a very brief period of time before achieving independence.

The other two structures which appear for the first time in some measure of specialized efficiency in the ferns are intimately connected with the vascular system. The first of these is the development of true roots, which supply water and nutrients directly to the vascular system. The other development was that of the leaves to which the vascular system carries the water. Leaves were not something new in the plant kingdom; mosses and liverworts both have leaves. The point is that in the mosses and the liverworts the leaves occurred on the gametophyte, the sexual generation: with the ferns and all plants above the ferns leaves occur on the sporophyte or non-sexual generation. This is the first instance of leaves occurring on a sporophyte.

Spores tend to appear on ferns under conditions that are relatively unfavourable for vegetative activity, usually when such activity is waning. As a result, the spores will either germinate immediately, or pass into a dormant phase before germinating. In either event they will germinate when conditions are unfavourable for vigorous vegetative activity. The result is a relatively small body. So, while the sporophyte has gained in body size, it has done so at the expense of the gametophyte.

This trend towards a diminished body size for the sexual generation and an increased body size for the sporophyte generation is one which will increase as higher and higher plants evolve, the difference in size between the two growing ever greater and greater.

It is worth looking at the life-cycle of a typical fern to see how it solved the other problem barring the way to progress. In the ferns, when a spore germinates it produces a tiny sexual generation gametophyte, called a prothallus. This is a tiny, flat body, much like a liverwort, for which it was for a long time mistaken, normally more or less heart-shaped. It lies flat on the ground, is green, and is self-sufficient. Instead of waving its sex organs around in mid-air, it produces them underneath itself. Which is a very sensible place to produce them when one bears in mind that the one place that can virtually be guaranteed to be damp on the land surface is when something, anything, is placed flat upon it. This seems to draw moisture to it. When one remembers that the sperm has to swim in order to reach the archegonium and fertilize the egg, one begins to understand just how sensible this arrangement is. They thus avoid the perils of dessication which face the sex organs of the liverworts and mosses: the result is that sexual reproduction becomes a pretty constant feature among the ferns, whereas all too often the mosses and liverworts had to revert to vegetative methods of reproduction because of the failure of their sexual systems.

In passing, it is worth noting that it is among the ferns that the sperm reaches its highest degree of specialized development as a sophisticated locomotive apparatus. The sperms of the ferns are objects of considerable beauty. The body (containing the nucleus) is unusually large for a sperm, and is spirally coiled, and has a very large cytoplasmic beak bearing an enormous number of tails, as many as forty or fifty. No doubt the increased number of cilia were necessary in view of the more bulky body: nonetheless, such a sperm must, in its day and age, have seemed a marvel of locomotive prowess.

Once the egg has been fertilized in its archegonium on the underside of the prothallus, it immediately starts to produce a vigorous non-sexual sporophyte generation, the roots and leaves emerging from beneath the shield of the prothallus. The sporophyte is thus parasitic on the gametophyte only for the very shortest period of time, just long enough for it to get its roots to moisture and its fronds into the light. In the process, of course, like most parasites, it destroys its host. It then grows vigorously, and at maturity produces spores which start the whole cycle in motion once more.

Having perfected this simple and efficient system, another element is gradually added, one which at first seems an unnecessary complication. This is the phenomenon known as heterospory.

The non-sexual, sporophyte, instead of producing innumerable identical spores, starts producing some which are larger than average, and others which are smaller than average. This difference increases until a point is reached where the sporophyte is producing both large spores and small spores, but no in-between spores. Each sporangium (that is the organ which actually produces the spores on the sporophyte) produces either a large number of small spores or a small number of large spores. These no longer produce a prothallus, all their energy requirements being pre-packed along with them.

This difference in size would be of merely passing interest, if it were merely a difference in size. But it is more than that; it involves a differentiation in function. The megaspores produce female gametophytes, while the microspores produce male gametophytes. The two have to get together for sexual union to take place. Thus in the alternation of generations while the spore-producing non-sexual individual remains a single, sexless thing, the sexual generation consists of two individuals, one male, the other female. While at this point in evolution such an arrangement seems merely an unnecessary complication, it was in fact the starting point of a characteristic common to all seed plants: indeed, without heterospory, seed plants would not have been possible. They evolved directly from the heterosporous situation.

In the act of becoming parasitic on the sporophyte the spores disappear from sight; they remain enclosed within the organs that produce them. The gametophytes have not merely become so reduced that they are no longer visible, they are scarcely there at all. However, they are still recognizable by external differences in the organs that produce them.

Whereas a spore is produced by a sporangium, a megaspore is produced by a megasporangium, while a microspore is produced by a microsporangium. The spores these two sexual organs produce become more and more unlike with the passing of the eons, and this quality of dissimilarity spreads to the organs which bear them. While an ordinary spore is produced by an ordinary sporangium on an ordinary leaf, the leaves producing megasporangia and microsporangia gradually become differentiated also. The specialized spore-producing leaves are called sporophylls. The sporophylls producing megaspores are modified to become megasporophylls, while those producing microspores are modified to become microsporophylls. Indeed, by this stage not only are the spores greatly differentiated, but also the sporophylls, which no longer bear very much resemblance to the normal leaves from which they have evolved. So great is the degree of modification that what one is faced with is an entirely different entity. It is, in fact, a primitive cone.

Once a cone had appeared, it seems inevitable that sooner or later the conifers would emerge. Yet evolution pursued some peculiar side-roads before the conifers were to reveal themselves and make a positive statement as to the next major evolutionary step in the sex lives of plants.

14
Pointers to a Missing Link

ONE OF THE GREAT PROBLEMS with the plant kingdom is that mankind is always trying to impose upon it a pattern that really is not there. In the human world it is perfectly possible to trace a man's genealogy from great-grandfather to grandfather, from grandfather to father, from father to son. That is a simple one-to-one relationship. Or a two-to-one relationship when one embraces in such a scheme the female ancestors also.. In arranging the plant kingdom into families we are trying to do the same thing: we are accustomed to saying that the grasses arose from lily forebears, and lilies from some ancestral monocot parentage, which itself sprang from some archetypal tree-like buttercup (the first flowering plant) and so on. But this is no longer a one-to-one relationship; nor even a two-to-one relationship. What one is looking at are whole communities, and communities do not have the clearly defined physical boundaries that individuals do. Because of this boundaries are blurred. And because boundaries are blurred, interpretation becomes a problem. Most of all it becomes a problem when there are gaps in the record. There is one such gap at a particularly interesting point in the sexual evolution of plants – a gap between those plants which bear spores, and those which bear seeds.

The problem arises because there are no living representatives of these fossil groups. So that although the fossilized sex organs can be examined in some detail by those who enjoy examining fossilized sex organs, there is no way of being absolutely certain of how these sex organs were used. A similar problem might arise a few million years hence were some jolly scientist to find the fossilized bodies of one man and one woman. He would probably reason that man and woman could not have reproduced at all because the limp, flaccid penis was too small and feeble an organ to penetrate the female passage, merely because he had no living material from which he could deduce that the penis could become erect under cardio-vascular constriction.

The problem we face in the plant kingdom is of a similar magnitude of triviality. What we do know is that, at the right period in time, a group known to us only from fossils existed. These were the Pteridosperms or seed ferns. Their fossilized sex organs turn up occasionally in lumps of coal, for they came into being towards the end of the Carboniferous period. The only question really is how they used these organs.

Like some of the higher ferns, they reproduced not merely by means of spores, but by microspores and megaspores. It is presumed that what happened is that the megaspore germinated while still attached to the parent plant. In so doing it formed a small prothallus which bulged outwards from a split in the spore casing. At this stage the archegonium – the female sex organ – is fully matured, and ready for the sperm invasion. All that is needed is for a microspore, or sperms from a microspore, to fall into the entrance of the waiting female organ to fertilize it. The sperms will then fertilize the egg at the base of the archegonium, and give rise to a new plant, growing on a prothallus which is still attached to the parent plant. The new plant will not fall free from the parent until it has reached some size.

All of which probably seems of no evolutionary importance so long as one looks on it as merely a failure on the part of the megaspore to detach itself. However, if one looks ahead to the flowering plants, one sees that microspores are still around, (though refined to the point at which they become what we all call pollen) and one realizes that it is pollen (microspore) that is transferred to the waiting female organ, then this accidental failure of the megaspore to detach itself can be seen as a fundamental and extremely necessary step to any further evolution in the sex life of plants.

For what, after all, is the difference between a megaspore which is fertilized while still attached to the parent plant, giving rise to a small new individual, and a seed, which is also fertilized while still attached to the parent plant? The differences are very small, and involve little more than the gradually greater and greater concealment of the megaspore and prothallus within more or less superfluous protective adaptations of leaves, together with an increasing period of dormancy.

Seed ferns, the missing link, are a group of plants that in all structural essentials of leaf, stem and roots, seem to be normal ferns, but differ from them in that instead of liberating all spores as modern ferns do, the female spores remain attached to the plant, and it is while still attached to the plant, that germination of the prothallus and fertilization of the egg take place, together with the beginnings of the growth of the new individual. It is but a small step from there to a true seed, a step which can still be observed in some living fossils.

15
The Sexual Performance of Living Fossils

THE CYCADS, by any standards, are a weird group of plants, so weird indeed that even botanists have been known on occasion to mistake them either for palms or for tree ferns, for they bear some resemblance to both groups, yet belong to neither.

Weird though their appearance may be, it is their sex lives that are the weirdest thing about them. Indeed, when one considers that some of them live for some 3,000 years before enjoying any sex at all, and then die of exhaustion after what must be one of the greatest orgasms in the vegetable kingdom, their sex lives can scarcely be considered anything other than extraordinary in the extreme.

The most striking thing about the cycads to the modern eye is, without doubt, their brilliantly coloured sexual organs.

But to what purpose, these garish colours on the sex organs of these fossil forefathers? Is the colouring there for its own sake, an end in itself? Is it the outcome of a superabundance of sexual *joie de vivre*? Or does it have some ulterior purpose? With human logic we assume that there must have been some purpose behind the circus colours of these sexual organs, but this need not necessarily be so. It is a point to which we shall return. It is tempting to see in the colouring of the cycad sex organs some sort of forerunner of the highly coloured flowers of the flowering plants – but they are still millions of years in the future.

The cycads are widely regarded not only as the forerunners of the conifers, but also as the forerunners of some of the flowering plants. Because there is a theory for which there seems to be much evidence, that the flowering plants did not evolve from the conifers, but that both groups evolved from the cycads, the conifers evolving slightly earlier than the flowering plants. Indeed, it seems likely that at least part of our confusion as to who arose from whom came about simply because we tend to make the fundamental assumption that in the plant kingdom there is a straight line drawn through evolution proving that the ferns descended from the mosses, and the mosses from the algae: what we overlook is that in the plant kingdom, as in our own families, there are often other offspring, the more or less coeval brothers and sisters. While in the genealogy of a family we can trace lateral movements of this type, it is much more difficult to trace this type of lateral movement in the plant kingdom.

The living cycads are a pretty diverse group of plants, and the fossil evidence suggests that in the past they were even more diverse. Just two examples from the now extinct *Benettiales* family show differences in sexual performance which in one generation seem to point the way to flowering plants, and in the other seem to point the way to the conifers.

The fossil genus *Williamsoniella* produces slender, regularly forked stems with narrow, entire leaves. It is a true vascular plant, with leaves, stems and roots. Where it differs from any plant before it is that in addition to the normal vegetative growth it also produces, in the forks of the stems, short fertile shoots, each bearing a cluster of sexual organs. This cluster looks superficially rather like a child's attempt to make a model of a large buttercup flower from the scales of some giant fir cone. That is not only what it looks like, it is also very much what it actually is. The 'petals' are made up of large, woody scales, very similar in structure and texture to those of the cones of pine trees, and at the base of each rounded, wooden, saucer-shaped scale there is, on the inner face, a sac containing a substance that is essentially pollen – a modified form of microspore. The central part of the structure round which these scales are arranged consists of a short, upright stem bearing ovules, an arrangement very similar to that found in the yews (*Taxus*) today. This short, upright stem carrying the ovules or eggs is simply a modification of that type of megaspore which germinates, ripening its sexual organs and waiting until it has been fertilized, and the fertilized egg has started to grow into a new plant before falling off the parent. What is new here is that instead of the male and female organs being borne at random on any old part of the plant, they are borne together in a structure which is specially arranged to bring the two organs into close proximity, thus increasing the chances of the sperm finding the egg before it dies of starvation: swimming can be a very tiring form of exercise. It takes no great leap of the imagination to move from there to something very like an archetypal flower. And maybe that really was the way things happened.

When we turn to the other fossil genus *Cycadeoidea* within the

Benettiales, we find similar materials being used in a somewhat different arrangement. This plant looked very much more like one of the living cycads than did the other, both in general structure and in its reproductive organs. These were borne on short-stalked stems hidden among the leaf-bases. It seems that some of these structures were entirely female, while others were hermaphrodite. It is the hermaphroditic structure that is of most interest. This was made up of wooden scales, similar to those found today in pine cones, arranged around a central column in a spiral fashion. Near the base of this structure there was a whorl of complex pollen-producing scales, while at the tip of the central column there were a large number of sterile scales, through which the ovules or eggs just peeped. When pollen (the male microspore) fell free from the organs producing it, the chances were that at least some of it would get stuck on the sticky sterile scales, and that the sperms would have little difficulty in swimming from there to fertilize the ovules. Such a structure points the way, not only towards both the male and female cones of modern conifers, but also to the general arrangement of the sexual organs in primitive flowering plants.

Apart from these amazing and prophetic advances in the structures holding the reproductive organs, some quite important advances were made by the sexual cells themselves. The sperms, for example, reach their highest degree of development in the cycads. The sperm of a cycad is shaped (rather unromantically) somewhat like a well-known turnip, with the cilia or swimming tails forming a tightly wound spiral round the lower part of the body. These cilia are the most efficient propulsive organs in the plant kingdom: no other sperms can match them for sheer propulsive power.

By contrast, the only real change to take place in the female organ is that it becomes even more firmly attached to the parent plant, so firmly indeed, that it seems to have taken a decision from which it will never be able to step back again.

It is at this point in time, where everything seems to be progressing so smoothly for the plant kingdom, that it suddenly becomes vulnerable to the influence of other external factors. Plants, at every stage of their evolution, had had to adapt themselves to changing conditions, the most radical being the move to land, with all its hazards of wind and weather, of landslides and volcanic eruptions: now it has to face a new hazard, the quite unexpected hazard of having its sexual organs eaten!

16

Insects get in on the Act

WE LIVE IN what we would like to think of as a sophisticated, civilized world. The truth is that we shield ourselves with only the thinnest veneer, a façade woven of a gossamer fabric, from the realities of this world. For the world that lurks beyond the boundaries of what we care to perceive is a world throbbing with sexuality, vibrating with violence, a livid, lurid world of killing and eating, of bloodshed and battles. And sooner or later it is inevitable that such a world must impinge upon the seemingly placid world of plants.

It seems probable, looking back down the corridors of evolution, that ever since there were plants there were also predators to prey upon them in one way or another. It goes on now and it probably always has gone on.

What is different about the predators that suddenly interrupt the till now seemingly smooth passage of the evolution of sex in plants is that they prey upon the sex organs themselves. It may seem an acquired taste, but it seems to be a taste which one group of highly successful creatures acquired at a very early stage of their evolution, and this group is the insects.

Most people are vaguely aware of the part played by bees in the pollination of many flowers. And some people are aware that in fact a very much wider range of insects gets in on the pollination process. But relatively few people seem to be aware of quite how early in their history insects become linked with plants and how intimate that relationship is. The stark truth is that the plant kingdom as we know it today could not have developed without the insects: and the insects could not have developed without the plants. The very intimacy of the relationship suggests a very ancient relationship.

The insects started to rise to prominence during the late Carboniferous and early Permian periods. This coincides with the prime of the ferns, the brief span of the seed ferns, and means that the insects as a group were already well established by the time the cycads rose to prominence.

So what was the relationship between plants and insects in those distant days? Did they come to marvel at these multi-coloured sex organs like ants at a circus? Or did they come to give these unwieldy organs (some of the cone-like sexual structures of living cycads are over a metre in length) a helping hand, and if so, in what direction? Whatever their involvement in those far-away days it was to have a profound effect, not only on the sex lives of the plants, but on the development of the insects themselves.

It seems most likely that insects, probably very large beetle-like creatures, were first attracted to the sex organs of the early cycads as a source of food. The insects are, and no doubt were in those days too, a highly practical group of creatures. They had no time to waste watching marvels: and certainly no energy to waste on pollinating other men's flowers. The insects came to feed on the reproductive processes of the plants, and what they ate were the pollen grains, the male microspores.

Yet the question of how the relationship between plants and insects first began cannot be considered by looking at the insects alone. It is also necessary to look at the plants of the period. And what is interesting when one does that, is that one finds that a very considerable degree of uncertainty exists as to whether the primitive cycads discussed in the last chapter were capable of completing their sexual union on their own and unaided, or whether the insects played a vital role in the transference of the pollen to the female organ.

So long as one sticks with the conventional view that the conifers descended from the cycads, and the flowering plants arose from some now lost link with the conifers, one has to conclude that since the conifers release their pollen and allow the wind to carry it safely to the receptive female organs, the forebears of the conifers must have done the same. It seems very much more probable in fact that the primitive cycads were perfectly well able to carry out the whole business of satisfactorily arranging for the sperm to swim under its own steam to the female organ, and then to penetrate it, so fertilizing it. Because if that were not the case, there is no way at all in which we can explain how it comes about that the sperms of these cycads have the most effective propulsion power units of any sperms in the vegetable kingdom. In nature, nothing elaborates itself unless there is some survival benefit in its so doing.

Obviously the first insect or two came out of curiosity. But they must have found out pretty quickly which parts of the sexual equipment of the cycads was best value, either in terms of flavour or nutrition, and then concentrated, in their hoards, upon that. But was it the male pollen grains upon which they made their meal, or the very much more nutritive female egg?

There is no way, at this distance in time, to know which they came for. Perhaps they came for both. Perhaps different insects came for different purposes, some to eat the pollen, others to eat the eggs. There are arguments to support both points of view.

One argument that very much favours their having come to eat the pollen is the abundance of pollen that some of these primitive cycads seem to have produced. They appear to have produced far more than was actually necessary for successful mating, unless they were very inept at the process. Probably, if the pollen was being eaten, those plants which produced the most pollen had the greatest chance of producing offspring and therefore were likeliest to survive. Generation after generation would have favoured ever greater and greater quantities of pollen being produced, until eventually the plants were producing a surplus, and it was this surplus upon which the insects fed. What supports this line of argument is that the conifers, which are believed to have descended from these early cycads, are wind-pollinated, a very chancy process requiring the production of vastly more pollen than can ever be used. This superabundance of pollen could well have arisen in response to the pollen grains of the forebears of the conifers eating the pollen.

The argument which most strongly supports the idea that the insects came to feed on the eggs, which are after all the most nutritive part of the plant, takes the view that, since the great majority of plants which have survived to tell the tale have developed carpels to protect the precious egg or eggs, that protection must have arisen in response to some depredation of the egg or eggs at some early date. Plainly, had the carpels not developed, the eggs would not have survived, so the species would not have survived.

There is no question of having to take one argument or the other. Both together satisfactorily explain the observable facts: neither explains those facts satisfactorily on its own.

The insects came, and they came to eat. And if, as they were scrambling around and over and among the plants' sexual organs, they happened to brush some pollen from the male organ to the female organ that was, to begin with, purely accidental.

There is, however, one theory which combines the best of both of the other two, and this is that what the insects really came for was not to eat the sperms, or the eggs, but to eat the fruits which resulted from the union of the two. On a cone-like sex organ a metre or more in length, seed was already ripe and ready for eating at one end while at the other the female organs were still awaiting the penetrating pollen grains. So that the insects which came to eat, in the process also pollinated.

The first immediate reaction of the plant kingdom to the invasion of their privacy by the insects was to try to repel them and avoid their interference. That route led, in the short term, to the conifers. The other option was to exploit the insects, to adapt in ways that would enable the plants to share their planet with the insects. And that, in the long term, was to give rise to the flowering plants and all the diversity of ways in which they coexist with the insects.

17
Male Supremacy

THE CONIFERS ARE, in a way, the silent legions of the plant kingdom: legions because, even without the interference of man who with military precision likes to plant them in straight lines marching across mountain-sides, they form vast, almost monospecific stands on their own, crowds gathered on mountain-sides, gazing silently down into the valleys below; silent because, if you stand in a forest of conifers, you will be standing in a world that is about as near to silent as you are ever likely to hear this side of the grave; pine-needles beneath your feet absorb the sound; the erect trunks deflect sound; the needles deaden it; at most you may hear the wind passing high above the tree tops. Silently they stand, straight, and silently they mate, asking only the wind's aid.

For quite suddenly the conifers abandon something that has been so much an integral part of the sex lives of nearly all the plants below them on the evolutionary ladder: something which had reached its highest perfection of development only one step back in time, with the cycads. They abandon the motile sperm.

But they did not do it all at once. Within the conifers there are two groups, the *Coniferales*, which are the conifers with which most of us are most familiar, and another, more primitive group, the *Gingoales*, of which the Ginkgo, the maidenhair tree, is the only representative with which we are likely to be familiar. In this group, the sperms still retain their primitive motility: the ginkgo is the only conifer we are ever likely to find in gardens of which one can truly say that the sperms still swim.

For the rest, the conifers have abandoned sperms with swimming tails. Indeed, the organization of the male organs has been highly modified, as indeed have the female organs, because the conifers chose to repel the insects, rather than to attract them. And this led to far-reaching changes, not only in their sex lives. The conifers, most of them, produce in their sap and in their needles, insecticides more subtle, more deadly, than anything produced by man in the way of insecticides. That is just one of the measures the conifers took to protect themselves from the emergent insects. The others were more intimate.

The most fundamental change was in the mode of getting the sperm to the egg. Since a sperm with a tail would have been highly vulnerable to the small insects, the tail was abandoned, and with it the swimming habit. But then a large microspore would also have been vulnerable to the insects, which meant that the conifers had also to evolve a very small microspore or pollen grain. It then had to face and overcome the problems of bringing the pollen grain close enough to the megaspore and its egg to ensure fertilization. In a land situation, and with a sperm that could not swim, the pollen grains had to be brought into very close proximity indeed with the eggs. So the conifers did what at first seems to have been rather a strange thing: they, quite literally, abandoned everything to the winds.

To do this successfully they had to accomplish two quite different things: in the first place they had to produce incredibly large numbers of microspores or pollen grains, since only a small number would have the remotest chance of reaching an egg. And they had to make the pollen grains light enough to float in the wind. And this they did by evolving two little sacs, each of which is filled with air – veritable buoyancy chambers which make the pollen grains virtually as light as air.

These pollen grains are produced in enormous numbers in collective organs known as cones. The male cones are produced in spring, clustered at the tips of the young branches, the cones themselves, on close examination, turning out to be merely modified leaf stems with modified leaves. Within this structure hundreds of stamens peep out, and on the underside of each stamen there are scrotal sacs containing the pollen. When these are ripe they burst, throwing the pollen to the winds.

The sheer quantity of pollen produced by the conifers is quite amazing. Country people are familiar with the sight, but people living in the supposedly civilized environs of a city have probably never seen the clouds of pollen blowing off the conifers. If you have never seen it, it is well worth while. Standing under a large pine tree on a clear, warm but windy day in spring, you can see the pollen coming off the trees in clouds, blown on the wind, drawn out from the head of the tree to form swirls and spirals on the vortex. With large specimens of some of the prostrate garden conifers, especially the junipers, it is possible to shake the whole

plant, and clouds of what many people mistakenly think of as dust, can be shaken out. So prodigious are the quantities of pollen produced by conifers, especially by whole forests of them, that it is scarcely to be wondered at that in times past, before we had any proper understanding of what pollen was, people used at times to be terrified when clouds of pollen came rolling down towards their villages from the pine forests on the hills above, like some visitation from above. Then there were the even stranger visitations of sulphur, no less, but it was merely the rain washing floating particles of pollen out of the skies, and washing everything yellow. Today all that most people know about these showers of pollen is what the pollen count tells them.

With the pollen released into the air, the organs which produced it wither, die and fall off the tree. Meanwhile the female organs are preparing themselves for the visitation. The female cone is positioned differently on the tree: it appears in place of a long shoot, so is placed at the outer edge of the tree, the most propitious to catch any passing pollen. The female cone looks quite different from the male cone: it is made up of row upon row of double scales, always firm, at first fleshy but later becoming wooden, usually brightly coloured, red or green. The scales are arranged, like the needles, in a spiral round a central core which is simply a modified stem, and they fit together very tightly, protecting their precious burden of eggs from the greedy, prying jaws of insects. The eggs are borne at the base of the scales, each scale producing two ovules, each of which contains a single egg. As the cones come, as it were, into season, they secrete a sticky resin which seeps down between the scales and eventually more or less covers the outer surface of the cone. The purpose of this secretion is to provide a sticky surface to catch any pollen grains which happen to be blowing by. It seems a haphazard method: obviously only a minute number of the pollen grains produced ever actually land on the female cones. And even the few that do still have to find some way to fertilize the well concealed egg. But what is worse is that they have to wait a year before they can do it.

For a whole year, the pollen grains sit on the sticky outer surface of the female cone, vulnerable to all sorts of insects and moulds, before they germinate. When finally the pollen grains do germinate, what happens is that they gradually grow a protrusion which penetrates the tightly closed scales. This is the pollen tube, a development of the prothallus of the microspore. As they grow, the pollen tubes carry the sperms with them, and once they have penetrated the egg, they release the sperms directly into the egg. It is by this curiously phallic outgrowth that the conifers overcome the problem of having sperms without tails.

A female cone is probably best thought of as an assemblage of megaspores, contained within modified megasphorophylls. Within it the ritual of the alternation of generations is still acted out. What actually happens when the sperm penetrates the egg is that a tiny plantlet begins to grow, living as a partial parasite on the female prothallus. Once it has incorporated all the foodstuffs available it stops growing and passes into a dormant phase. There it rests.

If one accepts as a definition of a seed that it is a plant in embryo, but dormant, then this is the first true seed on the face of the earth. For it will not germinate until the scales which protect it have opened, let the seed loose on the wind, and the wind has allowed the seed to fall to ground, where it will once again be wakened into activity by warmth and water, not to start growing, but to continue the growth which it had already started within the cone.

Thus, even with the conifers, the alternation of generations is still there, though concealed within the plants. The pollen grain is a microspore: the male cone an assemblage of microsporangia. The male prothallus, as it were, is the pollen tube, which conducts the sperm to its destination. Thus the sexual generation of the conifers passes almost unnoticed, largely living out its life hidden within the scales of the cone. But it is still very much there. Just as it is still there in the flowering plants, but even better concealed within the floral architecture.

18
The Final Flowering

TO THOSE OF US who are used to thinking of plants as green things that produce flowers, it must seem rather strange to have read through this far without encountering any flowers. Yet this is a proper perspective; flowering plants are not really very much older than the primates, both of which are extremely recent arrivals on this planet. There were plants of one sort or another around for about three and a half billion years before a single flower was produced. And if there seem to be a lot of flowers in the world today, that is partly an illusion created by those of us who cultivate plants for the beauty of their flowers, and partly a reality in as much as that the flowering plants are the dominant vegetation in the modern world, outnumbering individual for individual all other types of plant put together.

One of the many reasons for the success of the flowering plants is that they chose the other option open to them when the insects began invading their sexual organs and eating their offspring. While the conifers turned their backs on the insects, developing defences of various types, the flowering plants, or angiosperms to give them their proper name, choose to go along with the insects, and it proves to have been the better course after all. Their development and their diversity are as great as that of the insects with which they have chosen to live: the two kingdoms impinge on one another, interacting in a most intimate way. Just how intimate that relationship has proved is something at which we shall look rather more closely in the next part of this book. For the present we need to confine ourselves to just how the sexual organs of more primitive plants evolved into the rich variety of flower forms with which our world is so gloriously decorated.

Before explaining the specific changes that took place in the sexual organs and sexual cells of the plants that were to flower, we need first to glance back to the world of the living fossils, the world of cycads. Those curious plants, with their reptilian trunks, shared their world with the dinosaurs and pterodactyls. There

were no flowers in that world: the most colourful thing produced by the plant kingdom at that time was the multi-coloured cones of the cycads. As we have already seen, the cycads which bore seed heads that look like cones seem to have given rise to the conifers. It seems likely that those that had their sexual organs arranged in the way that a child might make a flower out of the scales of a giant fir cone did in fact give rise to flowers. The whole area is much disputed: learned experts write learned books on the subject, each presenting a differing view of the origins of flowering plants.

With our human passion for imposing order on other things, we tend to look for a single, archetypal flowering plant from which all other flowering plants sprang – some vast vegetable progenitor. It is most unlikely that there ever was one. It is far more likely and far more in keeping with modern knowledge to assume that flowering plants did not arise once and once only at one point on the globe and at one moment in time, but rather that they arose on many occasions in many places at different times and from different ancestors.

If we look for a missing link between those plants which flower and those which do not flower, we shall probably never find one, because there probably never was one. It is only because we tend to pigeon-hole things into their categories, neatly labelling one group flowering plants, and the other non-flowering plants, that we need to look for a missing link. In fact evolution was a more or less continual process, a gradual shading from organs that we could not call flowers to those which we do call flowers. So far as the sex life of the plants goes, the step from non-flowering to flowering was a very small one: many far more momentous steps had already been taken.

Most of what most people think of as flowers have little to do with the reproductive processes of the plants. They are simply flags that can be waved to attract the right insects to the right organs for pollination to take place. Usually these brightly coloured organs are petals, and as such they have evolved from leaves. It is the jewel displayed at the heart of the structure that is really interesting.

The jewel at the centre of every flower consists of some sort of arrangement of the sexual organs. What is perhaps more difficult to realize is that microsporangia and megasporangia are still present, usually both in the same flower. The essential parts of the flower are the pistil (or pistils) which is the female organ, a modified megasporangium, and the stamen (or stamens) which is a modified microsporangium. (It is surprising, really, how little has changed!) The rest is decoration, however purposeful that decoration may be. Petals, sepals, tepals, and all the other decorations stand to the central sexual organs of a flower in much the same relationship as a vagina to the uterus: it is fun, but not strictly necessary for reproduction.

Thus transmogrified, the alternation of generations goes on even within the flowering plants. The microsporangium is a stamen; the megasporangium a pistil. The microsporangium/stamen produces pollen grains: the megasporangium/pistil conceals the egg.

The male gametophyte is reduced to its lowest functional physiological form. It consists quite simply of three cells, two of which are sperms. The other forms the pollen tube which, as with the conifers, conducts the sperms to their destination deep in the womb of the flower.

The female gametophyte is similarly stripped of inessentials. It consists of merely a few cells, one of which is the egg.

The bare essentials of a flower, from a reproductive point of view, are the stamens, which produce the pollen, and the pistil, which excretes at its tip a sticky fluid to which the pollen grains adhere. The pollen grain grows a pollen tube, just as it does with the conifers, to conduct the sperms to the eggs. But instead of one sperm passing down the pollen tube to the egg waiting in the womb of the flower, two sperms are passed down. One will penetrate the egg, and their two nuclei will fuse: the other, and this is entirely new, will fuse with another cell which is not an egg: on fusion of this cell with the spare sperm, it produces a quantity of nutritive tissue called the endosperm.

At the same time the superficial tissues of the ovule develop into a hard coating – the seed case. A seed, therefore, represents three generations in one structure. The seed coat and the tissues immediately beneath it have been derived from the structures of the sporangium, and thus represent the older sporophyte: the endosperm represents the female gametophyte: while the embryo itself is the sporophyte of the next generation.

Once this is realized it becomes apparent that germination is properly the time at which the female egg-cell is fertilized by the sperm. After this it very rapidly forms the embryo sporophyte plant of the next generation. It then, normally, passes into a

dormant phase. What is generally regarded as the germination of a seed is in fact simply the arousing of the embryo from its dormant state. The new plant is already there: all that germination, in the sense in which it is usually used, achieves is the opportunity for the embryo to develop into a full grown sporophyte.

PART THREE
COPULATION

19
The Sexual Organs Displayed

A TYPICAL FLOWER, if indeed there really can be said to be such a thing when so many variations have been played upon a single theme, consists of four kinds of structures, the first two of which are seductive, the second two productive. These are: 1) the sepals, which together make up the calyx; 2) the petals, which together make up the corolla; 3) the stamens, which together make up the androecium and 4) the ovary or ovaries, which together make up the gynoecium. These are all produced together on the conical or thickened tip of the flower-bearing stem called the receptacle.

Take a typical and fairly primitive but at least well-known flower – the common buttercup. In this there are five green sepals, which enclosed and protected the flower while it was in bud. Then there are five bright yellow petals, at the base of each of which there is a tiny flap, which is the nectary. There is then a large number of stamens. Each stamen consists of two parts: there is a thin 'stalk' called a filament, and at the end of this there is an anther: it is the anther which releases the pollen. The stamen is therefore the male sexual organ standing in relation to the sex life of the plant in much the same way as the penis does to the sex life of most male mammals. At the centre of the flower are the carpels, each of which consists of a stigma with a sticky tip and an ovule at its base which, upon fertilization, will develop into a seed.

That, in essence, is the basic structure of a flower, and is recognizable as such in most flowers. The variations on this theme are endless, but they all mainly consists in alteration to the proportions of the flower, and even more in modifications to the seductive parts of the flower – the petals and so on. It is precisely by means of these changes in the floral structures that botanists arrange flowers into their families – though other factors besides merely the flower are taken into account. Some fairly basic changes which can occur are, for example, in the number of the parts: in the buttercup the basic number is five. There are five

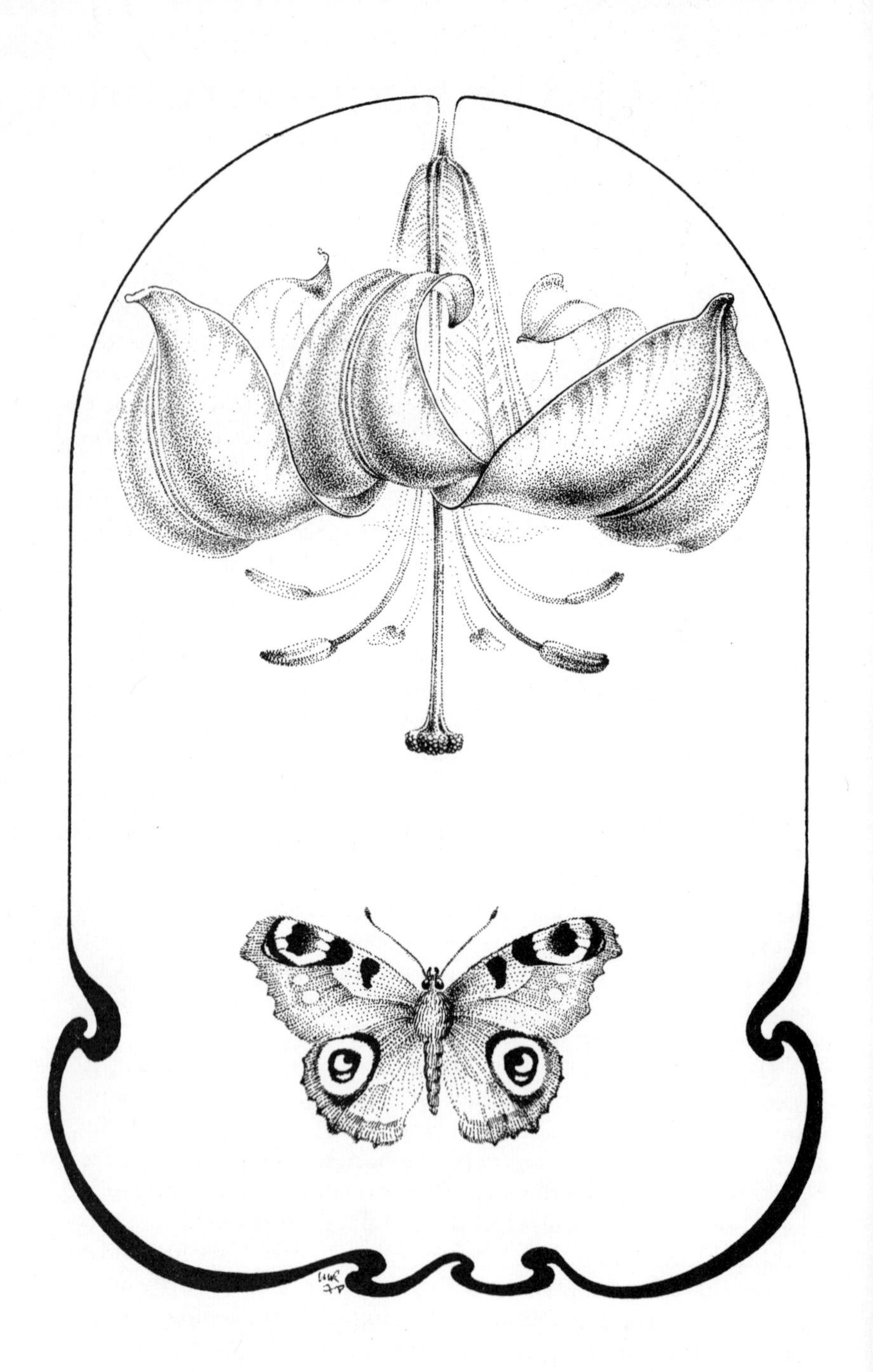

carpels and five petals. In the lily family and perhaps more obviously in the iris family, the flower parts are in threes: three carpels and three petals. Relatively few families have only a single carpel, but the grasses are one important exception. The number of stamens or carpels can vary even more than this. In some families the stamens are simply described as 'numerous'. The number of ovules within a carpel can vary considerably: most usually there is only a single ovule at the base of each carpel, but in some exceptional cases these too, may be 'numerous.' In one orchid, *Maxillaria*, the three carpels have been estimated to contain over one and three quarter million seeds – each of which developed from a single ovule.

While flowers are most typically hermaphrodite, they are not always so. Where male and female flowers are produced on the same plant, that plant is said to be monoecious. If they are born on separate individuals they are said to be dioecious.

Flowers may be produced singly or in groups. Where they are produced in groups there are a number of arrangements which they can assume. Where the flowers are produced round a single central stem, the arrangement is called a raceme, and the flowers at the base of the raceme open first. However, such an arrangement is properly only a raceme if each individual flower is produced on its own short stalk: if the flowers are stalkless or appear so, then the arrangement is called a spike. If the flower-producing stem ends in a single flower, but other stems produce other flowers on side branches off the stem, the arrangement is called a cyme. There is a further arrangement which is sometimes mistaken for a cyme but differs from it in that all the flowers occur on short stalks produced from the same point: such an arrangement is called an umbel. There are one or two further arrangements of flowers in groups, but these need not be discussed here. The point is merely that all these different arrangements can be used diagnostically when one is trying to determine the particular place in the plant kingdom of a particular flower, and that such arrangements have presumably arisen through evolutionary forces working on the plants and on the manner in which they are pollinated.

The structure of the individual flower as well as the various ways in which they are grouped together are all means used by the plants to display their sex organs in the most attracting and alluring possible ways, ways which are often extremely colourful. There is really no comparable behaviour among animals: true, some monkeys have brightly coloured behinds and one particular group is notable for the conspicuous turquoise testicles of the males, but these are exceptions, whereas flashy sex organs are the norm in the plant kingdom. The reason is simple. Animals can go in search of their mates: plants can't. They have to use sexual intermediaries, a sort of postal semen service, and to ensure that the postman calls they resort to some very extraordinary methods of seduction.

We tend to take it very much for granted that flowers produce pollen, just as we take it for granted that humans produce sperms. In fact the process is a complicated one. The stamen, the flower's equivalent of the penis, starts life as a tiny appendage on the receptacle. Gradually as this enlarges the different parts, the filament and the anthers become clearly distinguishable. The anther is the most important part of the stamen, for it is the anther which produces the pollen. The pollen itself is produced by what are called pollen mother cells, which themselves are formed by the division of the cells which form the inner wall of the anther. These mother cells undergo meiotic division, producing a tetrad of four cells: normally each of these cells then rounds off to become an individual pollen grain containing only half the number of chromosomes needed to produce a plant. Once these cells have been produced the walls of the anther begin to break down, so providing nourishment for the developing pollen grains. Finally, the pollen grain itself undergoes a curious division, forming two cells, a vegetative nucleus (which will grow into the pollen tube and penetrate the female organ) and a generative nucleus (which will fertilize the female egg).

Meanwhile, a similar process is going on in the ovule, the womb of the flower. The nucleus of one of the cells near the centre of the mass of cells in the ovule undergoes meiotic division, thereby producing four cells. The lowest of these enlarges to form the embryo sac, and in the process of enlarging crushes the other three cells out of shape. After which life gets more complicated. This enlarged cell itself divides into two and each of the two again divide into two, producing two groups, each of four nuclei. One nuclei from each group then comes together to form a single large nucleus at the centre of the embryo sac. This large cell is to the plant what egg-white is to an egg. Of the remaining nuclei one and one only enlarges to form the egg. The embryo sac enlarges and continues to enlarge digesting neighbouring cells, so that by the time the flower is ready to be fertilized it has come to form a relatively large, hollow chamber.

Thus the stage is set for the sexual act: the parts are ready for action. They have only one problem: they cannot, like animals, simply come together of their own volition. For this they are entirely dependent on highly specialized sexual aids.

20
Seduction

IN THIS WORLD there is very little point in having highly evolved beautifully designed and functionally efficient sexual organs unless male and female are able to meet somewhere.

In the animal kingdom this is all fairly easily arranged: mobility means that the animals can deliberately mate whenever and wherever the urge takes them.

The biological function of sex, whether in plants or animals, is to ensure the continuity of the species. In general, but not in some notable exceptions, two conditions must obtain before mating will take place. First there must be mutual attraction between the male and the female, and then the female must be sexually receptive. The first function of any flower is to seduce a reliable intermediary. The techniques of seduction employed by flowers are many and varied; beautiful, fragrant, occasionally bizarre.

We tend, being human, to look at flowers through human eyes. The fact that they generally please us, by their colour, shape or perfume, is a happy chance for us, but biologically unimportant to the flowers. The birds and bees quite literally, see them in a different light. Thus bees, for example, are good at seeing colours falling within the wavelengths which produce yellow, yellow-green, blue-green and blue. They seem to be unable to distinguish between reds and greys. On the other hand those flowers which are predominantly pollinated by birds normally fall within the orange-red to red range. It is probable that birds see the yellow, yellow-green, blue-green and blue colours merely as grey. Most of the flowers which are pollinated by nocturnal creatures, whether moths or bats, are white.

It is perhaps worth reminding ourselves about the nature of colour. Colours do not exist in themselves: they only exist because different pigmentary materials reflect light at different wavelengths. White light can be divided by passing it through a prism into its various colours, with violet at one end and red at the other end. The range of wavelengths visible to man fall between approximately 390 and 650 nm. Below the lowest figure is ultra-violet, which we cannot see, while above the highest figure is infra-red, which we cannot see either. The vision range of bees falls within the range 300 to about 550 rnm. Thus they are able to see ultra-violet, which we cannot, and they seem to have a particularly acute visual perception of ultra-violet when in combination with the other colours they can see. The mixture of ultra-violet and yellow, for example, produces a colour which we cannot perceive at all and which seems so far as we know to be specific to bees and is therefore known as 'bee-purple'. If colours from the opposite ends of the spectrum are seen together they produce white. But what looks like white to us differs greatly from what looks white to bees. For man, if red is removed from white light, he sees what is left as blue-green. If red is added back into the light, he sees white. Bees, however, see white light from which the ultra-violet has been removed as blue-green: if the ultra-violet is added back in they see what we could call 'bee-white.' Plainly, seeing things in a different light can make very real differences. And it is a real difference which flowers exploit quite shamelessly in their need to seduce sexual inter-mediaries.

When you or I look at the flower of the creeping cinquefoil (*Potentilla reptans*), the flower that we see is a strong, rich colour. What a bee sees when it looks at the same flower is 'bee-purple', not a yellow flower at all. The wild cherry (*Prunus avium*), is noted for its brief but brilliant burst of blossom in spring, the whole tree becoming a cloud of white. Or is it? Those blossoms which we see as brilliant white are perceived by bees as blue-green. The flowers of the stinking hellebore may be green to us, but they are yellow to bees. Ling, (*Calluna vulgaris*), which we regard as having purple flowers, in fact, as any bee will tell you, has blue flowers. Confusingly, where we see blue flowers on speedwell (*Veronica chamaedrys*) they see violet ones.

This is not done simply to cause confusion. It is done for the purely functional biological purpose of attracting the right pollinator to the right flower. The colour of every flower has evolved to attract only those creatures which can see the correct colour there: but there is even more to all this than meets the (human) eye.

Many flowers in which there is a moderate proportion of

ultra-violet reflectivity, show clearly defined patterns marked by areas from which ultra-violet light is not reflected. Such markings are invisible to us, since we cannot see ultra-violet light anyway. And that in itself is a further safeguard, ensuring that only the right pollinating creatures are seduced by the flower.

In a great many flowers the seductive mechanisms go beyond mere appearances. The scents which they emit are carefully tuned to seduce the right visitors. The perfumes produced by flowers vary from the very sweet through degrees of sourness to strongly pungent smells. There may even be, just as there are with colours, extremes at each end of the scent range beyond our powers of perception, but readily noticed by insects or other creatures attuned to them. Research, again into bees (presumably because, although they are as yet of no known military significance, their economic importance is substantial), has shown that bees will waver in their flight path, and then orient themselves to fly towards a floral scent source.

We perceive scents subjectively, as we do colours, within the range of our own smell-spectrum. There are certain scents that seem to have an almost universal appeal, the popular scents of many young people being typically very sweet, violet, 'orchid', even primrose. It is surprising just how many of the perfumes we use are named after flowers or the plants that produce them. While it is the sweet scents that seem to have greatest appeal to young people, older people generally prefer more musky perfumes, a change that goes hand in hand with a move away from sweet to savoury foods.

From what is known of the behaviour of insects and animals in relation to flower scent, it seems that flowers tend to give off smells which attract a fairly broad spectrum of creatures. In general plants which flower during the day tend to have relatively sweet smells, while those which flower at night tend to have rather musky smells, sometimes masked under a supersickliness.

There is, however, one type of smell emitted by plants which may be considered, in the terminology of the present day, to be a 'special case.' These are the plants which emit the odour of fetid flesh (the carrion flowers) and those which emit the closely related smell of the fecal matter of carnivores. It is only to be hoped, for the sake of insects which are seduced by these unpleasing aromas, that they do not perceive them in quite the same way that we do or with quite the same connotations: for it has been found that the smell given out by these flowers, and which some insects find so seductive is almost identical with the mating scent of the female of that insect species.

The combination of colour and scent are far more seductive than either on its own. Yet the purpose of seduction is to achieve an end: the painted, perfumed beauty must offer some reward to the sot she has besotted. For seduction to lead to the end which it is setting out to achieve it must offer more than merely the initial attention-getting excitement, and it is customary in western society for there to be a set-scene on the seduction stage intermediate between the moment when he notices her, and the moment when the seduction is consummated. In our society it is customary that this set-piece involve food, preferably rich, and reputedly aphrodisiac if possible. Strangely, flowers offer the same intermediate reward: food.

There are really only three foods which a flower can offer the bird, bee, butterfly or bat which it has seduced: nectar, pollen or petals. Of these nectar is the most important today, but looking far back into the evolution of flowers and the intimately correlated evolution of insects, it seems almost certain that in the very earliest of flowers it was the pollen upon which the insects, primarily beetles, fed. However, pollen is too precious for a flower to allow it to be indiscriminately eaten. It is as important in the survival of the species as semen is in humans. Apart from the fact that in some primitive precursors of flowering plants, such as the cycads, beetles do eat the pollen, the general lack of confirmatory evidence in the fossil record tends to suggest that those species which allowed their pollen to be eaten did not manage to reproduce in sufficient quantities to survive.

Beetles are indiscriminate eaters, the goats of the insect world, devouring whatever is devourable. One theory of the reason why plants evolved petals round their sexual organs is that this was to provide food for the beetles and thereby distract them from eating the pollen. There are other theories as to why plants developed petals which, though seductive, are not actually necessary for reproduction, any more than copulation is actually necessary to reproduction in our species, the job being done just as efficiently by artificial insemination or ovarian implants.

What is more certain is that flowering plants developed nectar

for the purpose firstly of feeding their visiting sexual intermediaries, and later for more devious purposes. The processes of natural selection operating through evolution tend to suppress any features of organs which are unnecessary to the survival of any species: contrariwise, those organs which are of real survival value tend to become more elaborate and more efficient. The fact that nectar is found in such an enormous number of plants suggests that it counts very highly among those attributes necessary for survival.

The importance of nectar cannot be underestimated, for once it had been 'invented', the future not only of flowering plants but also of the insects which pollinate them became intimately and irrevocably interdependent. Probably in the earliest flowers which produced nectar, the nectar was produced where it was wholly accessible to any errant insect that happened to be passing. Such a flower could be pollinated by virtually any old insect which happened to have a sweet tooth. Flowers and insects lived in a state of more or less total promiscuity. In the course of evolution flowers gradually concealed their nectar, making it more and more difficult for the insects to find. This change in the flowers necessitated a complementary change in certain anatomical details of the insects, these changes tending progressively to limit the number of insects capable of pollinating a particular flower. The deeper within its floral skirts the flower hid the nectar, the longer the mouth parts of the insects had to become to reach the nectar.

In a flower such as the giant hogweed (*Heracleum sphondylium*) or the sea spurge (*Euphorbia paralias*), in which the nectar is fully exposed, any passing insect can feed on it, beetles, flies, moths, butterflies or bees: not all of these will actually effect pollination, but they will all be seduced and get fed. Nectar that feeds creatures which do not effect pollination is wasted, so that some limitation of who shall get fed is an obvious development. As nectar becomes progressively concealed, so the number of visitors is reduced, and bees begin to play a more and more important part in pollination. In the great majority of flowers in which the nectar is completely concealed such as the Pasque flower (*Anemone pulsatilla*) or the raspberry (*Rubus idaeus*) bees are the only pollinators. If they had not evolved not merely their mouth parts but also their body shape and even their social structure, to fit in with changes in the flower, even they would not be able to reach the nectar of these flowers. In evolutionary terms increasing concealment of nectar ran parallel with increasing specialization on the part of the insects, and also to mutual selectivity.

There are many flowers in which pollen still provides the alluring food, but in those flowers in which pollen is eaten it is usually produced in the most enormous quantities, just as it would have been by the wind-pollinated ancestors of animal-pollinated plants. The problem that faced wind-pollinated flowers was that it was normal for male and female flowers to be borne separately, usually on separate plants. There was really no reason why an insect which had visited a male flower to eat the pollen, should visit a female flower which had no food to offer. In some way the female flower had to allure the pollinating insect to it, which is probably why flowers started producing nectar.

Plants, however, had one more wicked deception up their sleeve, though it is not a trick which seems to have gained sufficient popularity to become widespread. This is the production of pseudo-pollen, designed for the visitors to eat instead of the real thing. It is made of special nutritive tissues, rich in sugars and proteins, and both shaped and coloured to look like pollen. Usually the pseudo-pollen grains are attached to the stamens. So that the deception is complete.

There are still a few plants which allow their petals to be eaten by insects, usually beetles, and by bats. The great majority of these are relatively primitive plants, producing primitive flowers of the typical spiral form which so closely resembles a cone. If you examine carefully the flowers of a magnolia, you will often find that the petals have been eaten, and the same is true of water-lilies, again very primitive flowers, and the damage has usually been done by beetles. The problem, of course, with having all your petals eaten, is that once they have gone you will seduce no further lovers.

Very few, if any, flowers deliberately adopt all three love-food options: they tend to go for one and one only, and the favoured form is nectar. Yet even nectar is not a single, fixed, substance. It is a complex organic aqueous solution, made up almost entirely of sugars. Three sugars predominate: sucrose, (cane sugar), fructose and glucose. Sucrose is probably the most important of these. In chemical terms it is a di-saccharide which can be converted by the action of the enzyme invertase into equal parts of two mono-saccharides – glucose and fructose. This means that the flower trying to seduce an insect pollinator can ring the changes on the sugars it is offering, different flowers offering the nectar-mix most favoured by its likeliest pollinator. Three variations seem to be almost equally common: there are nectars in which sucrose predominates, there are nectars in which there are roughly equal parts of sucrose, fructose and glucose, and there are those in which fructose and glucose predominate. Specialization in nectar type is another means the flower has of ensuring that it seduces the right insect.

There are still further tricks a flower can get up to with its seductive nectar. An individual flower can vary the mix. The change tends to be away from sucrose to fructose and glucose, and since this change takes place over a period of time, it is almost certainly organized by the presence of the enzyme invertase in the nectar or the nectaries. In some plants male flowers have one type of nectar, the females another: thus the sallow (*Salix*) has male flowers whose nectar is sucrose-dominated, but female flowers are fructose-and glucose-dominated. But arrangements of that sort are strictly for gamblers: the bee that visits the male flower for sucrose, may not bother with the female flower because it does not like fructose and glucose together. Perhaps the cleverest arrangement of all is the device by which flowers manipulate not merely the rate at which nectar is produced, but also its sugar concentration.

In most plants there is a peak in nectar production and sugar concentration around mid-day, which just happens to be the time of day when the greatest number of insects is likely to be around to be seduced. For those who are sceptical of flowers having any sort of sense it is worth pointing out that this is highly unlikely to be mere coincidence. No plant in its right mind would put itself to the additional strain of running sugar and nectar producing machinery at full blast at mid-day if it were feasible to do it any other time of the day, since it is at mid-day that the plant is under maximum stress from water evaporation. Almost any other time of the day would be easier for the plants, but possibly not so effective.

Thus the three main seductive devices among flowers are colour, scent and food. They use both colour and scent to guide the visitors they have seduced to the food. Many flowers have specific guide-marks which the insects use to home in on the nectar in much the same way as an aircraft uses a radio beacon to home in on an airfield. The positioning of the nectaries ensures that the insect gets thoroughly dusted with pollen in its attempts to reach the nectar – and the better concealed the nectar is, the more likely the visitor is to get dusted.

There is little point in an insect getting itself covered in pollen unless it is going to transfer it to a suitably sticky and receptive female organ. A problem does arise: after all, most flowers have both male and female organs, so that the simplest thing would seem to be for the insect to deposit the pollen straight onto the style of the same flower. This however would not achieve one of the main functions of sex, which is the perpetual pooling and recombining of genes. Many plants have evolved essentially mechanical methods of preventing visiting insects pollinating them as well as receiving pollen. The thrum-eyed and pin-eyed primroses are classic examples of this sort of mechanical device: in the thrum-eyed flowers only the male stamens can be reached by the pollinating insect: in the pin-eyed flowers only the female style can be reached. Since there are roughly equal numbers of pin-eyed and thrum-eyed flowers, there is a fair chance of a pretty high fertility rate. Other methods are discussed in the chapter on contraception.

Plants seem to try to leave as little as possible to chance. No matter how efficient their mechanical contraceptive method may be, there is usually a fail-safe mechanism as well. Mechanical devices are designed to prevent self-pollination: the fail safe system is designed to prevent self-fertilization, which is a rather different matter. The system is incredibly simple – physiologically controlled self-incompatibility. In essence this works in a similar manner to that in humans by which antibodies rally round to reject foreign tissues, except that what the flower is doing is rejecting its own pollen.

Most flowers can so combine the various seductive assets with which they are endowed to attract specific pollinators or at least groups of pollinators. For example, in those flowers which offer pollen as the food-bait, the colours with which this attribute is combined are most likely to be red, yellow, or white. Such flowers are most likely to be visited by beetles, flies and short-tongued bees. In flowers in which nectar is the bait but the nectar is fully exposed, the colours may be red, pink, yellow, yellowish-green and all shades towards white, whereas in flowers in which the nectar is completely concealed the colours are likeliest to be red, blue or violet.

While flowers may not be able to flirt on quite so many seductive themes as we can, they certainly use great ingenuity in making the most of the limited means at their disposal.

21
Sexual Aids

MEN AND WOMEN, when they wish to mate, simply get together and get on with it, especially these days. For flowers life is not quite so simple. Since they cannot get together in the same way as men and women they have to make use of sexual aids or intermediaries to do the getting together for them.

While a relatively small number of plants uses the elements, such as wind and water, as their sexual aids, the great majority of flowering plants have to rely on the whims of an oddly assorted selection of creatures, including the proverbial birds and bees. It is worth looking, however briefly, at just what these creatures are, for as one does so it becomes very apparent that this strange relationship between plants and animals must have evolved very gradually through the ages, every step forward taken by a group from one kingdom being matched by a corresponding step forward by the group belonging to the other kingdom. Had they got out of step, the system would have got out of phase, and it is difficult to see how it could readily have been brought back into phase again. Indeed, it may well be that it is precisely where there seem to be links missing in the evolutionary lines that things may have got out of phase, so that there are no survivors from either kingdom to give us living proof of what went wrong.

The relationship between the insect world and the plant world is a very ancient one, as we have already seen. Possibly the relationship had its beginnings in those cycads to which beetles came as foraging visitors. The original purpose of their visits was presumably to gather food, and they probably ate the petals of sex organs of the plants they visited. Now evidently any creature which is stupid enough to nosh away indiscriminately on the floral equivalents of the penis and the vagina, is not going to be very much use as a pollinator. Yet all the evidence is that this is how the relationship started, and that much of what we see in the way of flowers today has arisen as a result of the flowers having deliberately to hide their all-precious inner sexual organs further and further within themselves.

The most primitive of the insect groups known to visit flowers today are still the beetles – which are the largest of all the insect orders. The order is sub-divided into two sub-orders, the smaller group of which contains all the predatory beetles, while the other contains 'the rest' – which in this case includes those which feed on or at flowers. The cockchafer and the rose-chafer, as well as many of the click-beetles, are very destructive of the flowers upon which they feed, and it is questionable whether the benefits they bestow by way of pollination offset the damage they do. Other beetles are less destructive. There is one that feeds on the pollen of one of the dogwoods, without doing any apparent harm, while another feeds on the orchid *Listera ovata* again without doing any discernible harm.

Flowers normally visited by beetles have had to make evolutionary adaptations to protect their ovules against the depredations of the beetles, which at best are indiscriminate feeders. They also emit a range of particular scents which seem attractive to beetles, either spicy as in the wintersweet (*Chimonanthus praecox*) or the crab (*Malus* spp.), or else a scent like fermenting fruit, such as that found in the allspices (*Calycanthus*).

The next largest order of insects involved in the pollination of flowers are the flies. These are readily recognizable in that they have only one pair of wings, the other pair being reduced to two short knobs known as balancers. Most flies are very small, and the flowers they visit are adapted to this. Since the flies have only very short mouth parts, the nectar tends to be fairly well exposed so that the flies can reach it. In general fly-pollinated flowers have an open, bowl-like structure. One special adaptation is found among the aroids (the arums and their relatives) in which the flies are attracted by a fetid stench, and are trapped inside the flower until they have completed the pollination work. Among the more familiar flies which visit flowers as pollinators are mosquitoes, midges and gnats.

Next in the evolutionary sequence come the butterflies and moths, with their showy wings and highly specialized mouthparts. Typically flowers pollinated by insects of this type have very heavy scents: common flowers include lilac (*Syringa vulgaris*),

honeysuckle (*Lonicera*), wallflowers (*Cheiranthus*) and carnations (*Dianthus caryophyllus*).

Among the less familiar, and indeed in some cases rather unexpected primitive, insects which pollinate flowers are a number well-known to gardeners as pests – things like the tiny thrips, which pollinate beets, ling (*Calluna*) and heaths (*Erica*), earwigs, and spring-tails, lacewings, leaf-hoppers and even the dreaded aphids.

Quite a large number of wasps are highly effective pollinators, but it is in the related bees that the art of pollination is brought to perfection. This is especially true of the social (as opposed to solitary) bees, and especially so of the genus *Bomus*, the bumble-bees. The honey-bees, *Apis mellifera* are, as might be expected, also highly effective pollinators, the more so since they have a longer season of active life than the other bees. Indeed, among the bees, especially the social bees, it is possible that the dependence upon the flower may have gone too far for the good of the flowers themselves. The social life of the bees, with all that this implies by way of communication, cooperative effort and even the distribution of labour within what might well appear to be a trades union organization, is such that they are able to exploit flowers in a way that no other group of creatures can rival. So great is the degree of dependence by the bees on the flowers that those flowers which are pollinated by bees do not merely have to provide either nectar or pollen, but usually both and usually in abundance. Furthermore, the flowers have to provide this feast for bees in sufficient abundance to ensure that the bees keep coming back to them.

Finally, there are some rather unexpected animals that have involved themselves in the pollination of flowers. There are rats and bats and numerous birds (especially in the tropics).

Probably the most unexpected and ludicrous pollinator of all is the elephant: but then the flower which it is reputed to pollinate is also the most ludicrous of flowers. This is *Rafflesiana arnoldia*, a plant of the tropics, parasitic on the roots of some *Cissus* species. It produces no leaves, but only, every now and then, huge flowers, sometimes more than a metre across and weighing as much as three hundredweight – so heavy that all it can do is sit on the ground. It is a huge lurid contraption, spotted and speckled with purples and maroons, and emitting the most powerful aroma of rotting flesh. Probably the idea that it is pollinated by elephants arose among the natives of Java because they could not think of anything else large enough to do the job. It is much more likely that it is pollinated by tiny flies or little beetles. But as Hilaire Belloc said, 'Let us never, never doubt/What nobody is sure about'.

22 Contraception

IF THE IDEA of contraception in flowers seems a rather odd one, that is merely because, as humans, we are applying purely human criteria to what we mean by contraception and what we hope to achieve by it. We might even learn something useful about ourselves were we to open up our minds and let the joys and sorrows of other species impinge upon our consciousness. The love life of a primrose or a petunia differs in a number of very real ways from our own *modus amore.* To understand their problems it is worth looking at our own a little more closely than we usually do. It is all too easy to accept contraception as one of the conveniences of the modern world, along with washing machines, disposable diapers, and throw-away plastic products.

Today mankind lives in a highly artificial manner, in a rather artificial world which he has largely manipulated to suit his own convenience. It is all too easy in a world in which too many of us tend to assume that food is something that grows in tins on supermarket shelves, and that condoms are something dispensed by machines in public lavatories, to forget what the real alternatives facing mankind are and always have been, at least since he became a relatively dominant species. So long as one accepts that the basic biological function of copulation is procreation, then the stark alternatives are either that every one breeds freely, but some of the offspring must starve, or else some kind of limit is put upon copulation so that the chances of conception are reduced. Primitive societies seem to have an instinctual understanding of the necessity of limiting their numbers, and also seem to know roughly what number is viable. They invariably operate checks of various types in order to keep the population down to the size which their territory can support. They do this by customs, often of religious or purportedly religious origin, mostly in the form of taboos: thus in many tribes any man who comes into contact with the blood of a menstruating woman is to be shunned by the whole tribe for a year or more. Since men dislike being isolated from society in this way, they shut away the menstruating women in a compound of their own. Other taboos are designed to prolong the period spent breastfeeding a child, since lactation is believed (somewhat erroneously) to prevent conception. Other taboos prohibit sexual congress during certain seasons of the year, certain phases of the moon, or alternatively only permit it at certain times approved by custom. Modern rhythm methods of avoiding conception are merely developments of such taboos. Taboos of this sort work up, to a point, but they usually have to be supported by more radical action to keep the population within its limits. Twins, triplets or deformed children are killed at birth: people are not permitted to grow old; once their usefulness to their society is over they are fed to the hyenas or left to perish on their own. Perhaps one of the strangest things of all is that there is strong evidence to suggest that in some primitive societies conception is avoided merely by the desire not to conceive: or conception is achieved only when there is a positive desire so to do. Modern man has lost this knack, but recent biochemical research provides ample evidence that one's thought processes can produce substantial changes in one's body chemistry. So it is possible.

In socially advanced but not necessarily technically advanced societies mechanical means are often employed to prevent contraception. A traditional method practised in India thousands of years ago was the insertion of dried cow-dung into the vagina: this formed a physical barrier to the sperms, which it also absorbed. In ancient Greece the dried corms of species of cyclamen were used much as a Dutch cap is used today. Sponges soaked in olive oil have been used from time immemorial right up till the present. More recently Johnson's Boswell, discussing his experiences as a randy young rake in London, mentions the problems of leaking seams in chamois leather sheaths.

Modern man, at least in the pill-taking Western world, turns to sex for consolation much as he turns to drink or drugs. The problems of unpremeditated conception have been largely conquered.

For plants the purpose of contraception is rather different. Plants go to a lot of trouble to display their sexual organs as attractively as possible: for them too, the prime biological function of sex is reproduction. But there is another aspect of sex which is of almost equal importance in plants, and that is that it is only through sexual union that the genes from two different parents can be pooled to produce a new individual which is not quite like either parent, thus ensuring that the species as a whole produces offspring varying in small but important ways, so giving them the capacity to adapt to the everchanging world in which they live. Of course, this exchange of genes is important in humans and other animals too, but flowers have a particular problem which seldom occurs in humans. That is that in the great majority of cases, they produce both male and female organs within the same floral structure: the arrangement is rather like that of some improbable human who had both a complete set of male reproductive organs and a complete set of female reproductive organs so arranged upon his or her anatomy that he or she could copulate with him or herself. Were the sex organs of humans thus arranged we should probably have to resort to some form of contraception in order to prevent self-conception, as it were, and that is exactly what many flowers have to do.

Self-fertilization nullifies the gene-exchange benefits inherent in sexual reproduction. It is therefore decidedly undesirable. The effects of in-breeding are well-known in both humans and other animal species. In-breeding imposes severe limits on the quantity as well as the quality of genes available for exchange, resulting in a weakening of the line. If this is true merely of in-breeding, it is very much more true where a line of plants evolves in which self-fertilization has become the norm. Charles Darwin, who was one of the most astute observers of natural phenomena, noted that 'Nature . . . abhors perpetual self-fertilization'. Normally plants only fertilize themselves as a last resort, when all else fails. But it does happen that groups of plants become isolated from their pollinators, and are forced into perpetual self-fertilization. Such plants have little evolutionary viability: they have lost that degree of variation bestowed by gene exchange which would enable them to survive in a changing world. They are doomed to extinction, however well they may appear to be surviving for the present.

Because of this many flowers have developed elaborate contraceptive means and mechanisms by which they prevent self-fertilization. However complex these mechanisms, there is, however, usually one rider, a fail-safe device that allows self-pollination if all else fails. In *Cobea scandens*, the cup-and-saucer vine, for example, a normally insect pollinated plant, the flower can pollinate itself in the last moments before the corolla falls off.

Probably the most well-known and well-documented example

of contraception in plants occurs in the common primrose (or perhaps not quite so common primrose) *Primula vulgaris*. It was Charles Darwin who made famous the contraceptive arrangement here, though probably few remember having the difference between pin-eyed and thrum-eyed primroses expressed in quite that way. In the pin-eyed primrose the stigma occupies the entrance to the flower-tube, the stamens being concealed half way down: in the thrum-eyed primrose it is the male stamens that occupy the entrance to the flower-tube, the stigma being concealed half way down the tube. Darwin very convincingly demonstrated that it is easier for an insect to effect cross-pollination where male and female flowers are distinct in this way, than it is for them to effect cross-pollination in flowers which are always the same. Plants bear either thrum-eyed or pin-eyed flowers, never both types on the same plant. The phenomenon is known as heterostyly (because of the male/female dominance or subservience of the styles). It would be less confusing to those of us who are not botanists if such flowers were referred to simply as heterosexual. The difference between the two flowers is very much of the same order as the difference between male and female in *Homo sapiens*: a man is, after all, only a woman whose femininity has been suppressed. The male organs are merely modifications of the female organs, the ovaries becoming the testes, the clitoris enlarging into the glans, the labia into the penis and so on, changes which are more a matter of emphasis than of a radical redesign.

The thrum or pin-eyed phenomenon does not occur only in the common primrose: it is also constant in other primroses including the cowslip, *Primula veris*, the oxslip, *P. elatio*, the bird's eye primrose, *P. farinosa* and several others. While it is a highly effective contraceptive device, there seem to be quite often isolated populations of primroses in which this contraceptive measure is not taken. Such populations are usually small, and isolated. In these populations both stigma and stamens are the same length: what is interesting is that these plants are self-fertile, with perhaps as few as five per cent being outbreeding. All of which suggests that the mechanical device of having stamens longer than stigma, or stigma longer than stamens, is only effective as a contraceptive measure when used in conjunction with some other factor, in this case self-incompatibility. It is rather the sort of situation that obtains when a diaphragm is used as a contraceptive: the back-up of a spermicide is also necessary.

The mechanical type of contraceptive measure taken by the primroses is also employed by a number of other plants, including the perennial flax, *Linum perenne*, buckwheat, *Fagopyrum esculentum* the water violet, *Hottonia palustria*, the bog bean, *Menyanthes trifoliata* and the fringed water lily, *Nympoides peltata*. All these plants exhibit what botanists would call dimorphic heterostyly – or in plainer English, two forms of complementary nature.

The purple loose-strife is the classic example of a plant which uses trimorphic heterostyly (three-forms of complementary nature) as a means of contraception. In this case there are three possible lengths for the stamens, the male organ, or penis, to put it plainly. They may be long, medium or short. The long ones produce large pollen grains: the medium ones produce medium pollen grains; and the short ones, as may be expected, produce small pollen grains. Which is very much as though a man with a large penis should produce large sperms, a man with a medium-sized penis should produce sperms of medium size, and a man with a short penis should produce small sperms. But mankind has enough problems without getting involved in complexes concerning the size of his/her organ or the products thereof.

Within the overall population of purple loose-strife there are roughly equal numbers of each type, the long, the medium and the short. However, there is considerable variation of the proportions within any particular loose-strife population confined to a specific locality. The total population appears to retain an equal apportionment of each type. The system works like this: long stamens produce large pollen grains which are only acceptable to female organs of the same length as the male organ: medium length male organs produce medium-sized pollen acceptable only to female organs of medium length. Inevitably therefore the small female organ has to accept small pollen grains from small male organs. A small female organ will not accept pollen from a large male organ: nor will a large female organ accept pollen from a small male organ. A medium-sized female organ will accept pollen from a large or medium male organ, but not from a small male organ. Thus in trimorphy any one individual can be pollinated by pollen from two-thirds of the remaining population. Were mankind equipped with organs of three different sizes, and the rest of the rules to apply to him as well, it would undoubtedly do much to reduce the world population. At the individual level there are already enough intimate personal tragedies without wishing upon mankind the problems that could devolve upon him were these conditions to obtain: a man could spend years courting a woman only to find out, too late, that she didn't fit. Though perhaps in this promiscuous age it would not much matter.

Another highly efficient method of contraception which appears to have evolved independently both in the plant kingdom and in the animal kingdom is the rhythm method. While there can be few situations quite so ridiculous as that of the beautiful young wench who keeps her ardent lover waiting in anguish while she takes the temperature of her vagina with a thermomoter so small that it seems at any moment in imminent danger of disappearing inside her, after which she has to compare the reading of the day with the graphs showing the temperature fluctuations of her menstrual cycle which decorate her boudoir walls, plants manage matters with much more dignity.

Take the avocado pear. Every individual flower on every individual avocado pear tree bears both male and female sex organs. A clever piece of timing prevents the tree from pollinating itself. For each and every flower on each and every tree will open twice and twice only, once during the morning, and once during the afternoon. On those trees which the planters call male, the pollen will be released from the male organs during the morning, but the female organs will be receptive in the afternoons, while on those trees which the planters call female it is the female organs which are receptive in the mornings, and the males which release their pollen in the afternoons. By this arrangement the trees can pollinate one another, but there is no way in which they can pollinate themselves. The planters are, of course, not strictly correct in calling the trees males and females, since both are both and both will bear fruit – provided only that both are growing in relative proximity. It has yet to be revealed to those of us who are merely human what it feels like to be male in the mornings and female in the afternoons, but genetic engineers, who are working on these things, assure us that it will not be too far in the future. Presumably the male in us will spend the morning effete with ardour, while the female in us will spend the afternoon fishing for the thermometer we left inside ourselves yesterday afternoon.

There are, of course, other and less complicated means by which plants ensure themselves against self-conception. The most obvious of these is to bear the male flowers on one plant and the female flowers on another plant. It is true that man and woman have been constructed along similar lines, but then man and woman have the advantage of mobility, enabling them to bring their sexual organs into close proximity, the one with the other. Trees, lacking such mobility, stand disadvantaged. They have secured total self-contraception at the cost of a severe limitation of genetic exchange possibilities. Though pollen may be very light, and it is possible to quote staggering figures as to the altitude at which pollen can be found, or the oceans it may traverse, such figures are somewhat academic, since pollen is, in general, short-lived. Even the most outbreeding of plants are likely to receive pollen only from very close neighbours, while trees are likely to have sex relations only with their next-door neighbours. And, which is worse, they may go on having to have them for two or three hundred years. While the placement of the different sexual organs on different individuals ensures, among animals, the highest possible rate of gene exchange, it produces one of the lowest and slowest rates of gene exchange among plants.

Yet plants have developed what must surely be one of the most sophisticated of all contraceptive devices, the phenomenon known as multiple allelomorph incompatibility, sometimes referred to as an antigen-antibody reaction. What this boils down to is that just as in humans antibodies rally round to reject the implantation of any foreign tissues such as lights, livers or kidneys so antibodies rally round in a flower to repel from the female organ pollen which has been produced by the male organ of the same flower. But in many plants matters have been taken a step further. What has been described so far is merely contra-self-conception. In many plants the same principle has been carried a stage further, and an antigen-antibody reaction may take place in the style to pollen from males belonging to a specific sub-group within the species. For example, recalling the contraceptive arrangements of the purple loose-strife, the direct parallel would be if a long style would accept pollen from a long stamen flower, but would produce the antigen-antibody reaction when pollen from a short-organed male alighted on it.

The nearest naturally occuring parallel in humans is when, due to the hormone changes that take place during pregnancy, a female who has already conceived cannot conceive yet again at the same time. And the highest degree of sophistication yet achieved by man is to produce a little pill which, if taken regularly, fools the female body into thinking it is pregnant so achieving an antibody reaction to an incoming tide of sperms.

Yet among plants the antibody reaction is the most primitive of contraceptive devices, predating by evolutionary eons the merely mechanical means.

23
Ejaculation

EJACULATION SEEMS TO BE such a very specifically animal sexual mechanism that it comes as something of a surprise to find that it also occurs in flowers. The whole process of ejaculation seems to be totally dependent on an intimate relationship between excitation of the nervous system and muscular activity, that at first glance it would seem that, if one is to accept that ejaculation also occurs in flowers, one must attribute to them too, not merely muscular activity, but also the presence of a nervous system which can be sexually excited.

One might have a better idea of what may be happening in plants if one has a proper understanding of what happens in animals: the most thoroughly documented animal in this respect is man. In man (embracing woman) the initiation of the mechanical piston-and-cylinder action that leads towards ejaculation is under the control of the voluntary nervous system. Thus one may begin or cease this activity at any moment along the line until a certain degree of excitation has been reached, after which the parasympathetic nervous system takes over. Now this part of the nervous system is not under voluntary control, so that once this point has been reached one cannot voluntarily discontinue the activity. What may seem at the time to be the pleasurable climax to a pleasurable activity is in fact merely a series of involuntary muscular spasms: ejaculation is caused by a series of such spasms occurring at intervals of 0.8 of a second, and these ejaculatory spasms cannot be voluntarily initiated nor voluntarily terminated. After all the fuss man makes about his sexual prowess, he may be surprised to learn that the nett product of all the heavy breathing and heavy petting is about two teaspoonfuls of semen. A pig, on the other hand, can pump about a pint into any submissive female. Perhaps we should not, after all, regard it as an insult to be called a pig.

Whether or not one accepts that plants have any sort of nervous system, there are certain points of similarity. If plants do have a nervous system at all, it would seem highly improbable that it is of the voluntary type: it would seem that they must have some type of parasympathetic nervous system, otherwise there would be insuperable difficulties in explaining how they manage to maintain themselves. Leaves breathe, transpire and photosynthesize: they open and close their stomata (leaf pores) in response to external conditions. Plants initiate flowering in response to complex patterns of temperature and day-length. And they do these things involuntarily; many are automatic processes: some are specific to specific conditions.

The basic processes of man's on-going biological functions are similarly involuntary. We breathe, we maintain a heart-beat, eat and excrete without having deliberately to think about it. Indeed, when one considers the difficulties that most people encounter when they are asked to rub their stomach with a circular motion with one hand while at the same time patting the top of their head with the other hand, it is probably just as well that we do not deliberately have to tell our hearts when to beat, because if we did have to we should probably forget also to tell our lungs to breathe or our intestines to contract when they should. We should probably finish up telling our hearts to breathe and our lungs to beat. Besides which we should not have time to think of other matters – such as sex.

Since both animals and plants had a similar origin in simple, single-celled organisms, it does at least seem likely that, in spite of the very obvious divergences in subsequent evolution, in both kingdoms the basic life support systems depend on some parallel mechanism. The mistake we make, firstly in attributing to plants a nervous system similar in kind to our own, is that it is most unlikely that plants would have such a nervous system. Whatever sort of nervous system they have, the probabilities are that it operates in fundamentally different ways to our own. Indeed, part of our problem would seem to be inherent in our language. If we attribute to plants a nervous system we are liable to make the mistake of looking for nerves (similar to those found in animals) in plants – and we are unlikely to find them.

There is only one reason for having any sort of sensitivity system, and that is the survival of the species for at least long enough to ensure reproduction. The fear, pain and flight mechanism in animals are not there primarily for the survival of the individual: only for the survival of the species. There is no reason to assume that matters should be otherwise disposed in plants. Whatever sensitivity system plants possess is there to ensure the survival of the species. And so plants, not having to elaborate their fear/pain/flight mechanisms, can devote the greater part of their sensitivity systems to enjoying their reproductive activities. Which leads to the curious conclusion that plants may well enjoy their sex lives a great deal more than we do. Sensations of pleasure in humans are, after all, only chemical changes. So there is no reason to denigrate the sexual pleasure felt by plants merely on the basis that it is only chemical. Since their whole metabolism is slower than ours, and reproduction in so many ways is more important than in animals, they may well be able to bask in their sexual joys far longer than we can.

Once we start to look at the mechanisms of ejaculation in flowers it will quickly be realized that normally each flower can only ejaculate once, as the result of a single spasm. In the human animal, both male and female, ejaculation may be prolonged over as many as fifteen spasms. But then man and woman have but a single sexual system apiece. If one regards each flower as a total sexual system within itself, which it is, then many plants have hundreds, indeed thousands, of sexual systems each of which is capable of a, presumably, satisfying sexual experience. As many as 3,000 flowers have been counted on a tree peony growing at the Royal Botanic Gardens, Kew, while on some of the arborescent magnolias as many as 5,000 flowers have been recorded. 5,000 orgasms a season is quite something, especially when one remembers that many plants may go on having 5,000 orgasms a year for 200 years or more: and when you remember that each flower typically contains both male and female sex organs, that is something in the region of 2,000,000 orgasms in a lifetime – which exceeds by far the wildest erotic dreams of man.

While most plants are content to accept their sex passively, there are a number that take positive and vigorous action to ensure satisfaction. There are two basic methods of achieving ejaculation in flowers: one is a quick flick of the stamens, flinging their pollen onto their unsuspecting visitor; the other is exploding stamens.

One of the most amazing of flowers from the point of view of the potency of its ejaculation is perhaps not surprisingly, one of the flamboyant tropical orchids, one of the Castaneas. This is a group of orchids which have developed, instead of the usual finger-like rostellum typical of orchids, a sort of rather rude two-fingered apparatus, which in fact are highly sensitive antennae and act as finely balanced triggers (in some species only one of the two fingers has this hair-spring trigger quality, the other being more or less inert). When one of these triggers is touched it sets off a sort of catapult mechanism which can fling the sticky pollinia disc (sticky side forward) at least a metre. This really qualifies as the world record ejaculation in the plant kingdom. However, the purpose of the exercise is not competitive: it is quite simply to make sure that the pollinia become firmly attached to the rather large, lumbering beetles and other creatures which come to eat the flower. It only flings its precious pollen three feet if it misses its legitimate target.

These trigger or catapult mechanisms are probably the most common of the methods by which flowers ejaculate. Indeed, one whole genus of Australian plants, the so-called trigger plants (*Stylidium*) get their name precisely because they have just this sort of mechanism, the trigger being quite easily sprung by any visiting insect.

Trigger devices seem to be relatively common in zygomorphic flowers (flowers, that is, which are bilaterally symmetrical instead of the more usual type which are radially symmetrical). A particularly effective ejaculation mechanism is found in alfalfa (*Medicago sativa*), and broom (*Sarothamnus scoparius*). In these plants the vigour of the ejaculation is almost explosive. The stamen tube is spirally coiled and held in tension between the keel petals. The tension is so finely balanced that an insect alighting on the flower releases the spring and the stamen flicks upwards showering the underside of the visitor with pollen. In gorse (*Ulex europaeus*) the mechanism is, in effect, reversed. The stamens and stamen tube, together with the style (the female organ) hold the keel petals in tension. Just before the flower opens, the stamens and stamen tube dehisce. The flowers are much visited by bumblebees, and these clumsy creatures, in trying to force their way into the flower, spring the trap, the keep petals burst apart and the bumbling bee goes head-first into the stamens.

In some plants the ejaculation mechanism is more than merely a catapult-like device. In some flowers the stamens actually explode, showering visitors with copious quantities of pollen. The mountain laurel (*Kalmia latifolia*) is one of this group: so are several members of the nettle family, and it is one of these which shows explosive ejaculation at its most potent. This is the so-called artillery plant (*Pilea muscosa*), which is quite often grown for a time by the unwary as a houseplant. The moment the stamens are even lightly touched they ejaculate showers of pollen, indeed so prolifically do they do this that if the plant is shaken great clouds of dust-like pollen will be emitted. In the absence of insect visitors or a good shaking, they will deposit vast quantities of pollen on any polished surface. Both cornflowers and knapweed have a similarly delicate ejaculation mechanism, but are probably the only two plants in which the same stamens can ejaculate on more than one occasion. The mechanics of this sort of dry ejaculation are simple in principle, but nonetheless, surprising when found in something assumed to be as simple as a flower. The anthers are borne on the top of a hollow tube, and as the style grows it gradually forces pollen out through the anther tube rather in the way in which liquid may be pushed out of the tip of a hypodermic syringe: the striking thing is that the pollen is not pushed out continuously, but ejaculated in small amounts only when stimulated by insect visitors.

Among some of the less showy orchids, *Epipactis* in particular, the ejaculation is of something somewhat more like seminal fluid. The insect pollinators are lured into the flower, which scarcely opens, by the copious quantities of nectar upon which they feed avidly, to a point just below the rostellum. At the very gentlest, most caressing stimulation from the visitor the rostellum exudes, with positively ejaculatory force, a drop of sticky liquid which securely cements the pollinia onto the visitor.

Of course it is quite impossible for us, with the limitations of our merely human senses and sensibilities to know for certain whether or not plants, and the sexual organs they produce, do or do not derive any pleasure from these experiences. But in view of their active participation in their sexual experiences, at least in these plants, it does seem possible.

24
Defloration

THE TERM 'DEFLORATION' always seems a rather odd one when applied to human virgins, but perhaps that is simply because we no longer have in our language any other floral terms denoting the female pudenda. The term is, however, a singularly apt one when applied to the manner in which certain flowers are pollinated.

Take the aspidistra, for example, that well-known denizen of dingy corners in Victorian homes, the so-called parlour palm or cast-iron plant. This plant, a member of the *Liliaceae*, does not bear large showy flowers on long stalks. Indeed, its flowers are in no way showy and certainly do not bear any of the refinement usually associated with flowers in the lily family. So insignificant are they that many people who have owned aspidistras for years have never noticed them flower, though most plants flower at least occasionally. And the flowers are very easily overlooked. Not only are they borne at soil level, where few people would think of looking for them, but they are also virtually the same colour as the soil – usually described in books as purplish, but a rather brownish purplish. So much are they borne at ground level that the petals have to force the soil aside for the flower to open at all. The flowers, when they are noticed, are cup-shaped, with rather blunt-tipped petals which spread out against the soil. Perhaps the most remarkable thing about them is their scent, for they smell distinctly of very bad breath.

With a repertoire of charms such as those, one would scarcely expect them to be pollinated by the beautiful humming birds, let alone by the less lovely bees. Carrion flies are a distinct possibility, but are not in fact the pollinators. Aspidistras are pollinated by snails. (Which may, when one thinks about it, be one of the reasons why one very seldom sees on plants grown in the home the bright orange-red fruits which follow the pollinated flowers.) As the snails crawl around, leaving their slime trails behind them, they chance upon the aspidistra flowers, which have a similarly slimy

texture, and set about eating the petals, which are thick, fleshy, and no doubt something of a gourmet item in a snail's diet. In the process of eating the petals, pollen adheres to the glutinous surface of the snail, and gets brushed off onto another flower. All that is left of the flower once a snail has made a meal of it, are the main sexual organs – the floral parts have been eaten. An aspidistra is, quite literally, deflowered at the climax of its sex life.

The aspidistra is not the only flower to be treated in this way. Its close relative, the omoto (*Rohdea japonica*) also has flowers which seem to be a gastronomic delight to gastropods. The plant is a very close relative of the aspidistra, also being a member of the *Liliaceae*, and bears a general resemblance to the aspidistra: though little known in Europe, several varieties are grown as houseplants in North America, while well over 5,000 varieties have been named by collectors in Japan. The omoto bears its flowers on a short spike, produced at ground level. These spikes consist of greyish-purplish, very fleshy perianth segments massed tightly together in a manner which makes them look very like the densely produced leaves, but their presence is revealed, as with the aspidistra, by the odour of halitosis. Slugs and snails crawl among the leaves and over the flower spikes, devouring the fleshy perianth segments but apparently not touching the stigmas, yet collecting sufficient pollen on their slimy bellies to effectively pollinate the flowers as they go. The greedy gastropods, having deflowered one omoto, move on to the next, thus ensuring cross-pollination.

It is not only among the *Liliaeceae* that deflowering seems to be an integral part of the sexual process. The phenomenon is also well established for the Hawaiian screwpine, *Freycinetia arborea*, a plant of most curious aspect, with sword-shaped leaves like those of a yucca or phormium, and conical flowers very similar in general appearance to those of the Mexican breadfruit plant, *Monstera deliciosa*. The conical flower is surrounded by a number of fleshy, orange-red bracts (which are leaves modified to look rather like petals and to serve a similar attractive function). Although the flower evolved its curious aspect to be bird-pollinated, it is in fact pollinated by rats, which were introduced to the Hawaiian islands by mariners from Europe. These eat the fleshy bracts, which apparently have a pleasant rather sweet flavour, and in the process collect pollen on their fur and whiskers, pollinating the flowers as they move from one head to another. There are other screwpines which are deflowered in a similar manner by flying foxes.

In Australia the Banksias are frequently deflowered by little creatures known as dibblers – tiny marsupial mice – which are highly adept at climbing and whose long tongues are well-suited to reaching the nectar hidden at the base of the bristles of the bottle-brush flowers of this genus. However, the dibblers do not stop at the nectar: they also eat other parts of the flowers.

It seems rather doubtful that any of these plants deliberately evolved in such a way as to need to be deflowered in the sexual act. It is much more likely that they evolved in a manner calculated to entice pollination by some relatively common pollinating creature, such as beetles among the gastropod-pollinated aspidistra and omoto, or the bats in the case of the Hawaiian screwpine, and that these deflowering pollinators came on the scene later. Were defloration a highly successful mode of ensuring pollination, it would be much more widespread than it is.

There is one quite extraordinary and almost certainly coincidental point of interest among those flowers which are deflowered in the course of pollination and many which are not. It has already been noted that many flowers change their scent once pollination has been effected, or change colour, in either case the change warns the pollinating insects not to waste their time or energies on that flower. Where a flower is deflowered in the process of pollination, a very clear visual indication is given. It is as well, perhaps, that the same clear indication is not given in the case of humans.

25
Living Together

PERHAPS THE STRANGEST of all the relationships between flowers and their pollinators is that relationship which botanists call 'mutualism'. This is a very specific type of one-to-one relationship, where the survival of a plant species is intimately interrelated with the survival of one particular insect species. Seemingly, if the one were to perish, the other would also perish. True, one-to-one relationships do exist among other plants, notably among some orchids, but they at least leave themselves the possibility of survival even if their pollinating insect no longer visits them. An example is the bee orchid which in Britain has lost its natural pollinator: having found no other, it merely pollinates itself. The sort of relationship we are looking at in this chapter is far more specific. It is a one-to-one relationship in which the whole life-cycle of the insect is so intimately tied up with that of the flower which it pollinates, and the flower has specialized itself to such a degree, that both have become too specialized, to dependent on each other for either of them to survive should the other perish.

That there was something peculiar about the fruiting of the fig (*Ficus carica*) was known as long ago as the times of classical antiquity. Theophrastus recorded the custom, already of considerable antiquity in his time, of hanging a branch from a wild fig among the branches of a fig in the garden to ensure the ripening of the fruit. While he was a sufficiently acute observer to know that unless this was done the fruits would fall off the tree without maturing, he was not aware of the complexities of the reasons why this was necessary. Indeed, this practice is still carried out in some parts of the world where the modern hybrid self-fertilizing figs are not known, and the practice is known as caprification, presumably because the wild fig, known as the caprifig, has a strong goat-like odour. It may not be mere coincidence that the goat has been noted through the ages as a most lusty beast. The unanswerable question is whether the wild fig got its name because it performed its sexual function like a goat, or because it smelt like a goat.

Whatever the ancients thought about the pollination of the fig, the reality is far more complex than they could possibly have imagined. For one thing, they probably thought that the fig fruited but never flowered. Indeed. it is most unlikely that they ever even saw the flowers of the fig, for these are borne on the inner surface of a fleshy, hollow receptacle whose only entrace is not merely minute, but is closed by flexible scales. What you eat, when you eat a fig, is not a fruit in any conventional sense of the word: it is·a complete inflorescence that you are eating.

There is another thing that was most likely unknown in classical times, and that is that the fig does not bear one type of flower, nor even two types of flowers, but three types of flowers – male, female and neuter. This in itself is a pretty unique adaptation to its pollinator, but even stranger is the way in which each type of flower is produced during a different season, and is apparently very precisely adapted to the needs of the pollinating insect at each season.

The first of the three types of flower is produced during the winter, and contains many neuter flowers, and a smaller number of male flowers confined to the region of the entrance to the receptacle. Flowers of this type are invaded by tiny female chalcid-wasps which lay their eggs in the neuter flowers and then die – without leaving the flower. The offspring of these wasps complete their development within these neuter flowers, one wasp to each flower. The male wasps hatch first, and emerge from the ovaries into the inside of the receptacle. With a highly developed sense of purpose they then seek out the ovaries which are still occupied by females, bore their way into them, fertilize the females in a short but fatiguing orgy, after which they die. (The receptacle is by this time becoming a positive cemetery, with two generations of dead within it). The fertilized females then emerge from the receptacle, brushing past the male flowers which have only just opened and getting dusted in pollen as they squeeze through the narrow exit.

By this time it is early summer, and the second type of flower has formed. In this the receptacle contains either neuter and female flowers, or only female flowers. The new generation of wasps penetrate these receptacles, and lay their eggs in the flowers. However, only those eggs which have been laid in the neuter flowers develop. The mechanism which ensures this is very cunning. The neuter flowers in this generation of receptacles are female flowers modified in a highly specific way: the ovary of the neuter flower cannot produce seed, and the style is very short, leaving an open canal leading directly to the ovary. The female of the species of wasp which pollinates these flowers has a very long ovipositor, sufficiently long to reach the ovary of the neuter flower (which should perhaps more properly be regarded as a bogus female flower). Only those eggs which are laid directly onto an ovary will develop. In the true female flowers the style is long, blocking the canal which the ovipositor must penetrate in order to lay eggs on the ovaries. However, the wasps try to lay their eggs in the true female flowers and in so doing pollinate them with pollen brought from the flowers they have just left. Having done this, the wasps, as you may by now have come to expect, die inside the receptacles. Development of the next generation of tiny wasps takes place exactly as it did in the previous flowers, with the males fertilizing the females, then dying. The fertilized females emerge in the autumn, to take up residence in the third type of flower, which is much smaller than the other two types, and contains only neuter flowers. These are the overwintering quarters of the wasps, yet here again the whole cycle of death and rebirth occurs all over again. In spring a new generation of fertilized females emerges to begin the cycle all over again.

Even that, with all its complexity, is not the whole story. The male fig-wasps are highly modified creatures. Since they never leave the safety of the receptacle, they never have to fly: their sole purpose is to fertilize the females, and they have been modified through evolution to perform this function and this function only. They lack wings, have greatly reduced legs and very small eyes. The females are fairly normal, but they do tend to get their wings torn off in their struggles to escape through the narrow exit from the receptacle. There is, however, one very specific adaptation in the female. This is that when she lays an egg in a flower, she also secretes a tiny droplet of a special liquid which causes the ovary of the neuter flower to develop into a gall, which causes the ovary to enlarge, thereby providing food for the young fig-wasps.

Yet perhaps the most macabre part of the whole fig-wasp story is the way in which the figs serve as both womb and tomb for their pollinators. The males live out their entire lives inside a single fig receptacle, being born inside the receptacle and dying inside it. The females are born in one receptacle, and die in another.

Such a complex, intimate and mutually dependent life-style can only have come about through eons of mutual development: there is no stage at which the fig-wasps could have evolved faster than the figs, or the other way about. Both must have developed hand in hand down the ages. It is certainly one of the strangest marriages between plant and animal known. What is perhaps even stranger is that this has not only happened once, but many times, for in each different species of fig, pollination is secured by a different species of fig-wasp. So specific is this relationship that no hybrid figs are known to have occurred in the wild.

For such a relationship to develop each side of the partnership must benefit. The benefit for the wasps is that they have a secure breeding ground: the benefit for the figs, is that there is always a population of fig-wasps on hand to ensure pollination. Which is fine so long as it lasts: for one of the most noticeable conclusions that can be drawn from what we know of evolution, is that the more specialized a species becomes, the less able it is to adapt to an ever-changing world. The short-term future of species as interdependent as this seems assured, but the long-term future exceptionally uncertain.

The only other remotely comparable instance of a marriage between the plant and the animal kingdoms is that which exists between certain species of yucca and yucca-moths. In general there is only a one-to-one relationship to the west of the Rockies, all the species to the east of the Rockies being pollinated by a single species of moth, the so-called yucca-moth.

Both flower and moth have made considerable adaptions to each other: they are indeed, as one might say of the bride and groom, perfect for each other. One of the strangest of the adaptations made by the yucca flower is in its nectaries. In most species these are still present, and still produce nectar. The problem is that the yucca-moth does not feed from the time it is born as a moth, till it dies. The question therefore arises as to just who it is who is to benefit from the nectar. It seems that what has happened is that whereas in most flowers the nectar is there to attract the pollinating insect to exactly the right place on the flower to effect pollination, in the yucca it is there deliberately to guide any visitors other than its chosen moth away from the stigmas.

It is the female of the yucca-moth who plays the most important role here, just as it is the female that is most important in the case of the fig-wasp. Her behaviour on arrival at a flower follows in every case such a precise and stereotyped pattern that this can only have been acquired over millions of years. On arrival at a flower the female yucca-moth climbs up a stamen and bends her head very closely over the top of the stamen, at the same time uncoiling her tongue. All the pollen on the tip of the stamen is then gathered together and rolled into a tight, compact ball, and held fast under the moth's head. The moth may climb as many as four stamens, and collect the pollen from each in exactly the same way. Fully laden, she then sets off for another flower. On alighting at the next flower she then examines the ovaries with great care, to make sure that they have not already been visited by another female. If they have she keeps on searching until she finds a flower which has not already been visited. If the flower is right for her, she climbs up the stamens again, and then goes through between them onto the ovary. She then reverses down between the stamens and the ovary, bores a tiny hole, and lays a single egg. After laying the egg she then climbs up to the stigmas, which are united to form a tube, and very carefully and very deliberately works a small amount of the pollen she has brought with her, down the stigmatic tube, thus ensuring pollination. She then reverses down again, lays another egg, climbs back up, pushes more pollen into the tube and so on. She does this for each and every individual egg she lays. On average the moth will lay three eggs in any one flower, but occasionally she may lay as many as twelve. The great care to which the female moth goes to ensure pollination is not merely a neurotic mother's compulsive mania, it is actually very necessary, since an unpollinated flower dies very quickly. In ensuring the fertilization of the flower the moth is ensuring a supply of food for her offspring. Those ovules into which the moth has laid her eggs grow abnormally large. The remainder, of which there are usually plenty, develop normally. The abnormal cells provide food for the yucca-moth young. The normal ones turn into seed. Interestingly, seed ripens on the normal ovaries at the same time as the larvae of the yucca-moth emerge from the abnormal one. The yucca-moth larvae crawl down the flower stem to the ground, and pupate in the soil surrounding the plant. Adult yucca-moths emerge over three seasons so that, should the flowers fail to appear in some years, as they sometimes do, there will still be plenty of yucca-moths to fertilize the flowers a year or even two years later.

As with the fig-wasp, there must obviously be some benefit for both plants in this relationship. What the yucca-moth gains is a safe breeding place: what the flower gains is a dependable pollinator.

There is just one slightly disturbing footnote to the tale of this otherwise perfect marriage. There is a bogus yucca-moth, which also lays its eggs in the ovaries of the yucca flower, but does not bother to collect the pollen from one flower, or to ensure the fertilization of another flower.

The really puzzling question about such seemingly perfect relationships, is just how did they begin? What seems most probable is that it all began way back in evolution with the larvae feeding on the internal parts of the flowers. At that stage the moths were probably only occasional visitors. Gradually they came to place the eggs in the flowers to be sure that the larvae would have food. The flower would seem to have adapted to encourage this.

It is perhaps disappointing for mankind, who would so passionately like to believe in altruism, if only because he seems to be incapable of it himself, to discover that even those marriages which occur in the natural world seem to be based on the common, selfish desire to survive.

26
Drunken Lovers

SHAKESPEARE has one of the characters in *Macbeth* remark that while alcohol increases the desire, it impairs the performance. It would seem from the context that he was referring to the proximate fumbling, rather than the longer term effects so popularly and precisely known as drinker's droop. It is in the first sense that many flowers, and orchids in particular, affect their pollinating partners. What is most inhospitable in these flowers, is that it would appear to be their deliberate intent to intoxicate their guests.

Probably the most spectacular of these inebriating plants is the tropical orchid *Coryanthes macrantha.* It has an absolutely massive flower, several centimetres across, and the lower lip, the pouch, has become an extraordinary sort of bucket: so close is the resemblance that this orchid is often called the bucket orchid. This bucket is suspended at right angles to the rear of the flower, and moreover has a sort of spout. Just above this spout is the finger-like rostellum, bent at a right angle in the middle, and bearing at its tip the stigma and pollinia. The bucket is partially filled by a liquid secreted by two knobs at the base of the column. All of which may sound a very Heath Robinson sort of arrangement, but is in fact a highly sophisticated piece of pollinating engineering. For male bees of the genus *Eulaema* are intensely attracted by the flamboyantly colourful and strongly scented flowers. This scent attracts the bees not merely to the flower, but to a precise part of the flower, an area of special tissue which the bees scratch with their forelegs in order to collect the liquid scent. This liquid scent, however, affects special sense organs on the bees' front tarsi, and they rapidly become exceedingly drunk, in consequence of which they gradually lose their grip and fall straight into the aforementioned bucket, which has no doubt been provided for that purpose. Once in the bucket the drunken bees swim around on the surface of the liquid until they are sober enough to begin to work out how to get out of their predicament. The main problem is that the sides of the bucket are very steep, virtually vertical, and exceedingly slippery. The only way out is through the spout, and sooner or later most bees discover this. Even when the bee has finally realized that there is only one way out of the flower, its problems are not over: the flower simply does not let the bee escape that easily. As it crawls up the spout the bee finds itself in a fix, because the phallic, finger-thick rostellum holds it firmly between thorax and abdomen. The pressure the bee exerts upon this organ in its efforts to escape, ruptures the sacs which contain the pollinia. Thus when the bee does eventually depart, it has the pollinia firmly plastered on it. The first bee to visit a virgin flower really does have a rough time: it can take the bee as much as thirty minutes to escape from the bucket. Subsequent bees have a much easier time, partly because the sacs containing the pollinia have been ruptured, and the rostellum exerts much less pressure on these later visitors. Hopefully, these later visitors have already been plastered by a previous flower, in which case they will leave the pollinia from an earlier visit to a different flower on the now exposed stigma.

All of which may seem quite extraordinary lengths to go to simply to get the sperms from one flower to the vulva of another. It is almost as though some lunatic lover were to insist that his mistress, having first showered herself with expensive perfume and drunk a gallon of mulled ale, should walk a tightrope which has been carefully greased, should fall from those dizzy heights into a bubble bath pie from which she can only escape by crawling through an S-bend which is spring loaded with a device which slaps each buttock with a custard pie before allowing her to escape. Humans do in fact go through processes just as elaborate in the arousal stages that precede copulation, but the processes are in the main minute changes in the tone of voice, the size of the pupil of the eye, the colouring of the lips and so on: there is even, of course, the proverbial heavy breathing. But then humans mate directly one with another – at least that is the conventional norm. Flowers mate vicariously, through the agency of untrained and untamed go-betweens. Since such go-betweens are notoriously unreliable, the flowers are justified in going to almost any lengths to ensure that the go-between does in fact deliver the package. The end, in this case, fully justifies the means.

The bucket orchid is not the only member of its genus to resort to intoxicating the go-between bees. In the very similar

Coryanthes speciosa the pollinating bee is partially intoxicated by the scent of the flower at close quarters: it falls into the waiting bucket without even nibbling at the flesh of the flower. As it approaches the flower two droplets of a soapy liquid secreted by special knobs become stuck to the wings, effectively making the bee incapable of keeping itself airborne. It therefore plummets straight into the bath, without even touching the tightrope. From there on the process of pollination is the same as for the bucket orchid. One has, somehow, the distinct visual impression of a steady procession of groggy wet-winged bees emerging from these flowers, like harlots in a Hollywood rainstorm, or dirty old men in dirty old macs from some downtown dive of ill-repute.

One point of more than passing interest is that in *Coryanthes speciosa* there appears to be a shut-off mechanism. If a pollinating bee remains in the bucket for more than a certain length of time (the time period has been observed as being about forty-five minutes) the scent which attracts and intoxicates the bees ceases to be emitted. The next day the scent is there again, as powerful as ever. The presumption here is that the shut-off mechanism is a device to prevent self pollination. It would be all too easy for a drunken bee, emerging from the spout with pollinia on its body, to fall, quite literally, into the same trap again, thereby depositing upon the stigma of the orchid pollinia which it had received from the same flower. All of which suggests that the flower is in some way able to 'know' how long a particular bee has been in its bucket. Which once again begs the question of how it knows this.

The two orchids mentioned so far are not the only ones which intoxicate their pollinating visitors. The phenomenon is also found in species of *Gongora*, *Stanhopea* and *Catasetum*, orchids which are also notable for their ability to ejaculate. Among the most interesting of these intoxicating flowers are the swan orchids, which not only make their visitors drunk, but also in so doing force them to perform the most curious contortions. In one of the swan orchids, *Cycnoches lehmannii*, the flower is inverted. This is readily achieved in many flowers, of which the common violets are the most obvious example, by a curvature of the stem which presents the flower upside-down whereas were the stem straight the flower would be presented in the normal manner. In presenting its flower upside-down this particular orchid presents its visitor with a problem – a problem which the bee solves by entering the flower upside-down. The bee, therefore, alights upside-down on the lip of the flower: once there it is lured further into the flower in its efforts to reach the source of the intoxicating scent: as it approaches the source of the scent a projection on the lip forces the bee to let go with its hind legs in consequence of which its body swings down (like that of a trapeze artist) and, if the flower is a male one, touches the anther-cover. That triggers a discharge mechanism which very neatly deposits the pollinia in precisely the right place on the abdomen of the bee. Female flowers of this species are generally smaller, but they operate a very similar system, with the flower inverted and the bee having to enter it upside-down. The important difference is that the bee, in performing its trapeze act, swings past two hooks on the column of the flowers, which catch the pollinia so carefully positioned by the male flower. Pollination in this orchid seems to depend upon a degree of acrobatic skill which it is doubtful that the bees could perform when sober.

There are several variations on this theme. There is, for example, another swan orchid, *Cycnoches egertonianun*, in which the female flower is pollinated in the manner described above, while the male flower has a slender lip delicately hinged in such a way that, when a bee alights upon it, it bends down under the weight of the insect, thus bringing it into contact with the anther. In both *Stanhopea* and *Gongora* the drunken bees fall from the lip in such a way that their backs touch the column and the pollinia become deposited on the thorax. In *Stanhopea*, which has massive, waxy-looking flowers, there are two prongs which guide the bee with great precision as it falls. The size of the gap between the prongs determines the genus of insect that will effect the pollination. The majority are pollinated by bees belonging to the *Euglossa*, but those with a larger gap between the prongs are pollinated by the larger *Eulaema* bee.

Another orchid which prefers its clients to be inebriated is *Catasetum*, which has huge inverted flowers, the usually dependent pouch being an inverted sort of hood which the bees enter in search of the source of the intoxicating scent. The rostellum of these orchids has been elaborated into two antennae of considerable sensitivity which function as triggers. Sooner or later the drunken bee or other large insect visitor is bound to touch one of these triggers. The result can really only be described as an ejaculation, for the flower catapults the whole pollen-mass, sticky disc foremost, straight onto the visitor. The ejaculation is a powerful one: the mechanism can fling the pollen mass a full metre from the flower. In this case there is no apparent reason (apparent, that is, to anthropocentric humans), as to why the bee needs to be inebriated for the mechanism to work. It could well be that the intoxicant is merely an attractant, and the intoxication purely incidental. It has been established over several other intoxicating orchids that the bees appear to enjoy the sensations of intoxication, returning again and again to the source flower. Nor do the visiting insects appear to suffer any harm from their apparent alcoholism: both *Eulaema* and *Eglossa*, the two main genera of bees which visit intoxicating orchids, live for as much as six months (which is a relatively long time for a bee).

If the idea of a plant intoxicating its pollinating visitors seems an outlandishly curious one, it is worth remembering that pollination partnerships can only work and continue to work where there is a benefit for both partners in the arrangement. Curiously enough, it is not at all clearly understood exactly why bees enjoy getting drunk. It is no good drawing parallels with man here. Man after all is old enough to know what he is about when he gets drunk, wise enought to know better, and still foolish enough to go ahead anyway. Bees lack the type of ratiocination that leads to both wisdom and folly. Their behaviour is instinctual. Perhaps what is really significant is that in all the cases known in which flowers intoxicate their visitors, it is only the males that visit the flowers. This suggests that the intoxicating substances play a part in the mating behaviour of the bees. One theory is that the bees use an oil derived from the liquor to mark out their territories: another is that it is used to attract bees of both sexes to one spot. The marking out of territory with scent is a sufficiently well-known phenomenon: many animals do it, either with scent glands or with their urine. What is certain is that the bees enjoy the liquor, returning again and again to suitable flowers to collect and store the substance in spongy tissue pouches on their legs.

The explanation of why the flowers find it necessary thus to intoxicate their clients is even more curious. The reason is quite simply that bees, especially the big ones, make bungling, incompetent and clumsy lovers: brusque and business-like no doubt, but that is not always what is required of a lover. The flowers, by getting the bees drunk, are able to manipulate the bees to their own ends. The very idea that flowers – those beautiful, innocent things – would stoop to such depths of depravity as first of all to deliberately get their visitors drunk and then to use them to satisfy their own peculiar sexual appetites, might seem to come from some far-fetched science-fiction novel. In fact, it is simple fact. But manipulation is more than merely a matter of manual dexterity: it implies also a certain degree of intent or purpose. The intention is plain enough: it is to secure cross-pollination. and all that is being manipulated by the plants is the sexual behaviour of the bees: bees that go through pseudo-copulation with orchid flowers are also being manipulated, and having their sexual desires cruelly exploited.

Yet one is left with the slightly disconcerting impression that flowers that can manipulate large insects to the degree which the flowers discussed here manage to achieve, must be capable of some sort of feeling in order to achieve their own ends by methods at once so complex and so comic.

27
The Harlot Hoards

THE POINT has been made again and again that the successful pollination of any species of flower depends upon a partnership in which both partners benefit more or less equally. There is, as it were, a system of barter: a favour is given for a favour received: payment is made for a service rendered. The service is the pollination of the flower: the payment is normally made in kind. Flowers offer the birds and bees which pollinate them nectar, pollen or even fleshy petals which they can eat: some plants even go so far as to produce special protein rich pseudo-pollen which is provided specifically for the pollinating birds to feed on while dusting themselves with the true sexual pollen. Most of these devices are most highly developed in the orchids. Yet there is one group of orchids which at first glance appears to offer the visiting insect nothing at all. It offers no nectar, no protein-rich pollen, or pseudo-pollen, no aromatic oils, no intoxicating liquors, no edible parts. Yet is is successfully pollinated and regularly visited by bees and other insects. So it must offer something. The offer is of that most intangible of all things – satisfaction: or perhaps no more than merely the semblance of satisfaction.

This group of orchids is the *Ophrys* tribe, the bee orchids and their kith and kin of Mediterranean regions, the *Cryptostylis* orchids of Australia and the South American *Paragymnomma* orchids and possibly one or two others.

The similarity between the bee orchid and its pollinating bee was noted long ago: the similarities between some of its relatives and their pollinators was known as long ago as the times of Classical antiquity. Yet, curiously, the situation was for a long time totally misunderstood. The Victorians – often such acute observers, yet such appalling moralists – concluded quite erroneously that the reason why a bee orchid looks like a bee is to frighten all other creatures away. Had their deductive logic been as acute as their observational powers they would rapidly have realized that this would have left the orchids with no pollinators at all. It seems

that it took a total breakdown in the moral fibre of mankind for it to be realized that the sole rationale for a bee orchid looking like a bee is to attract a bee. That is, to put it plainly, to attract it in the way in which a female bee might attract it. And with much the same view in mind. For the stark truth is that the male bees try to mate with the orchid flowers.

The deception is one of the wickedest tricks in the world. These flowers are worse than harlots, lower than whores, yet they have much in common with these painted ladies. A whore makes her money by deliberately exploiting the sexual weakness of men. And in exactly the same way the bee orchid exploits the sexual weakness of the male bee. For among those bees which are the specific pollinators of bee orchids, it is normal for the males to emerge from pupation well before the females. They are thus unable properly to satisfy their sexual drives and, like the sailor home from the seas, have little alternative but to accept the services of the most accessible whore.

The trick is a clever one, for the degree of similarity between the flower of the orchid and the female bee is quite remarkable. It is the correct shape, the correct colouring, and, perhaps most important of all, it emits the correct smell. Any deficiencies in design are cleverly compensated for. The flowers, for example, have reflective patches which not only fool the bee into thinking it is visiting another bee when it is only chasing its own reflection, but also act as homing beacons like those on airfields, which the bees can use to home in on. The flowers have shiny spots which do look quite remarkably like eyes, and greatly reduced petals which resemble antennae: frequently the side lobes of the flowers have an appearance which makes them resemble the folded wings of an insect. But for the scent emitted by these flowers, one might dismiss the whole thing as playing at charades. But the scent seems to more than make up for any design faults: so closely does it resemble that of the female of the species, and so powerful an attractant is it, that bees will try to find these flowers even when they have been hidden inside paper bags. Once the bee has homed in on the flower of the bee orchid and landed on the lip, it finds that the lip has the right curves, the right protrusions and even the right degree of hairiness to fool it into thinking it has found a female. And the poor silly bee immediately tries to mate with the flower, an activity which it carries out at length and with vigour, in consequence of which it effectively pollinates the flower.

It would seem, however, that the bee is not completely fooled by the prostitute flower. It does not actually copulate with the flower, though it comes very close to it. The bee is sufficiently excited by the flower for it to go through the movements associated with copulation in bees, sufficiently excited for the male organ to be extruded and rubbed back and forth against the flower. The essential difference between this performance and true mating is that in this performance the bee does not ejaculate. There are other minor differences between this and the real thing. The mating movements are carried out for longer than in true copulation, and usually also more violently. With true anthropocentric sensitivity one can almost feel for the frustration of the bee as it gradually dawns on it that it has been duped and this is not, after all, the real thing; in its frustration it will quite often viciously bite the lip. I believe Catullus said something pertinent about that.

Biologists distinguish the mating motions which a bee goes through with a flower of this type from the real thing by calling it pseudo-copulation. In spite of the fact that the bees get, what in public school slang would be called a dry rub, they persevere, moving on from one flower to another, as though living in perpetual hope of encountering a genuine female. Of course in the fullness of time they do, but by then they are either worn out by their exertions or highly trained performers, depending on one's point of view.

While the bee orchid is certainly the most famous of the orchids pollinated in this way, there are numerous others, nearly all belonging to the *Ophrys:* most attract bees, but a few attract flies. One species, *Ophrys speculum*, is peculiar in that the bees which pollinate it insert themselves, as it were, backwards into the flower. There it alights, sitting astride the lip, tail up, head down, and the tip of the rostellum suspended just above its head. The bee then plunges the tip of its abdomen into a fringe of long red hairs at the end of the lip, and goes through a series of rapid, trembling and almost convulsive movements, in the course of which the pollinia become firmly attached to its head. The movements are undoubtedly copulatory movements, and the female bee is remarkable for the fringe of red hairs on her body. These red hairs guide the male when copulating with the female, which it does in the normal position: it seems that the orchid can only procure pollination when the bee is in the head-down position.

That pseudo-copulation works as a mode of pollination is plain enough. In one of the few genera which have been documented fully, it has been found that as many as forty per cent of the capsules contain fertile seed, which compares favourably with figures for other methods of insect pollination.

While there are undoubted problems in explaining the evolutionary origins of these orchids, it can only be assumed that they have evolved through natural selection from some archetypal orchid in which the attractants were scent and nectar. Scent obviously plays a very important role in the orchids in their present form. The other adaptations must have taken place not only very gradually but also in parallel, with the bees adapting to the orchids at the same time as the orchids were adapting to the bees.

Orchids, and especially the orchids in this group, have often been called the prostitutes of the plant world, but the epithet is questionably apt. True, prostitutes exploit the sexual weaknesses of the opposite sex, traditionally the female of the species exploiting the male. And undoubtedly the orchids are exploiting the sexual weakness of the males they attract. But a prostitute is in the game purely for the money: she does not gain sexual satisfaction from the encounters from which she profits; her only and dubious satisfaction is that of exploiting her clients. You do not exploit those you love: you only exploit those you despise. Modern psychology suggests that prostitutes generally hate and/or despise men, and in exploiting them are getting their own back on them by degrading them. It would seem that the tender relationship between an orchid and a bee is of a somewhat different nature. It is much more as though the bee were exploiting the orchid in the way in which a nymphomaniac exploits her lovers, for a nymphomaniac is never satisfied, just as the bee is never satisfied. Yet in the relationship between the bee and the orchid the attraction is mutual: the desire reciprocated. The one species could not survive without the participation of the other.

The idea that the flower is exploiting the bee suggests that the flower is capable of some form of intent or purpose. If one is not prepared to accept that the flower can have intent or purpose, the alternative is to look on the relationship in very much the light in which one would look upon the relationship between a high-powered executive with a frigid wife and the self-lubricating inflatable rubber dolly bird he bought through the pages of a glossy girlie magazine. All the evidence points away from the idea of the flower as totally passive and inert. It seems to participate, be it in a passive female role, in the copulatory activities of the bee, and there is no reason to believe that the flower does not enjoy the union just as much as the bee. After all, pollination is to a flower what copulation is to you or me, and did the flower not possess some sensibilities it would not be able to know when pollination had taken place. There are many means by which flowers show that they do know when they have been pollinated, such as the shutting down of the scent-producing mechanisms, or changes in the colour of the flower.

It would seem fair to suggest that, in a competitive world in which both the animal and vegetable kingdoms depend for their survival upon sexual union, these orchids have taken a logical line of development to its logical conclusion.

28
Bondage

FROM TIME TO TIME one comes across lurid accounts of various human sexual perversions, one of the most common of which is bondage: it seems that some people cannot gain full sexual satisfaction unless their partner has been firmly lashed by wrists and ankles to the bedposts, and so can offer no resistance to the advances and demands made upon them, or if they do try to resist, their struggles only serve to further excite their partner. Such peculiar sexual mores in humans are easily explained away by sociologists, anthropologists, psychologists and all sorts of other – ologists who say that this sort of thing is quite normal in a decadent society and so on, and the same thing happened in Rome in its decline. What is perhaps surprising is to find that there is a large number of plants which can only procure sexual satisfaction by bondage too.

Probably the best-known family in which bondage is a prerequisite of sexual satisfaction is the *Araceae*, of which the genus *Arum* is probably the most familiar. Indeed, the cuckoo plant or lords and ladies (*Arum maculatum*) possesses one of the most remarkable of all pollination mechanisms. The strange, curious, not to say weird inflorescence is produced on a short stalk a few centimetres above the ground in late spring/early summer. The inflorescence is made up of two parts each of which, because they are so different from the floral structures of other plants have special names. The sexual part of the flower consists of a central column or shaft called a spadix, and this is protected by a large, leafy hood called a spathe. It is the spathe which is the showy part of the flower, and in some members of the arum family such as the arum lily it is large, white and very showy. The sexual organs of the plant are confined to the spadix, which in most aroids is extraordinarily phallic in shape. The female flowers are produced all around the base of the spadix and are reduced to their sexually functional minimum: they are merely ovaries topped by stigmas. Above them are a few sterile flowers which, in spite of their impotence, do play a part in the bondage mechanisms of the plant, while above these are the fertile male flowers, which again are reduced to their sexually functional minimum: they are nothing more than short-stalked stamens. The uppermost flowers are again male, again sterile, but have long hair-like appendages. Above these sexual organs proper the spadix is extended into a more or less club-shaped phallic structure, usually a lurid purplish colour. The showy, leafy spathe is constricted just below this swollen, club-like phallus, so that the sexual organs are enclosed in a narrow-necked chamber. Insects are lured to the inflorescence by the combination of spathe and spadix, and enter the sexual chamber, the copulatorium, as it were, through a narrow space between spathe and spadix. Once the insects have entered the chamber they are trapped, at least during the first, female stage of the flower's development. The more they struggle to escape from this bondage, the more they rub themselves against the receptive female stigmas, dusting onto them any pollen they may have picked up from a previous harrowing experience of the same sort. Once the female flowers have passed their period of receptivity, it is the turn of the male flowers, which emit a relatively large amount of pollen. It is only when the trapped insects have endured this shower of pollen long enough that the spathe allows them to escape. The next arum that the insects visit will have the pollen from the first flower dusted onto the stigmas during the first stage of their bondage, thus effectively ensuring cross-pollination. In the process it seems that some of the insects are foolish enough to struggle so much that they perish in the attempt to escape.

While the broad outline of the sex life of the common arum (*Arum maculatum*) is well known, there has been no in-depth study of it as there has been of the Mediterranean arum (*Arum nigrum*). On this a veritable Kinsey report has been written. It is somewhat larger than the British *Arum maculatum*, and it is possibly this which makes it easier to study. Indeed, a blow-by-blow account of every move could be given. The spathe opens during the night. Throughout the following day the upper, phallic part of the spadix emits a revolting fecal odour, which attracts hoards of dung-frequenting flies and beetles. These generally use either the phallic knob of the spadix or the inner surface of the spathe as a landing platform. Now, both the phallus and the inner surface of the spathe are covered with special cells, producing not only a smooth surface but also a copious number of downward directed conical papiliae

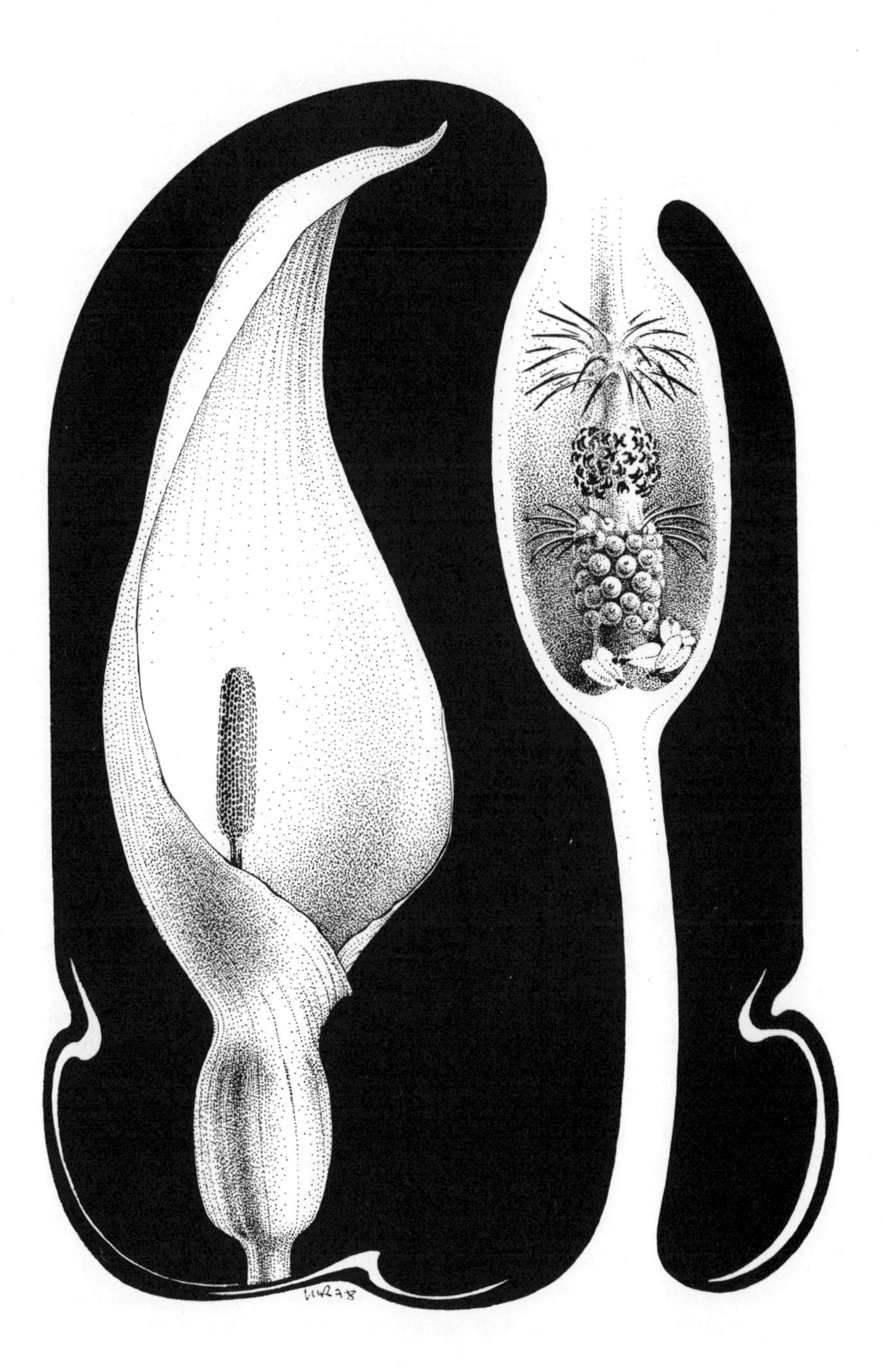

which provide no foothold for the visitors, an effect which is heightened by the emission of an oily secretion. The insects therefore fall, both spathe and spadix combining to direct the insects where they want them to go. However, before the insects reach the copulating chamber they encounter a ring of sterile male flowers, whose function it is to sort the men from the boys. It does this in a way similar to the manner of a gravel grader: only the chosen, smaller species pass through: larger flies and beetles get no further and are free to fly away. Those that pass through finish up in the copulatorium. If the insects have already previously visited a similar flower they will have been dusted with pollen, and this will come off on the sticky, receptive stigmas as the insects struggle to escape. Once the female flowers have been pollinated, the pollen tubes grow with exceeding rapidity, and the ovaries are fertilized in an extraordinarily short time. Once this has happened the stigmas wither, and as soon as that has happened the male organs are ready to ejaculate, which they do with vigour, showering the insects with clouds of pollen. Since the male organs cannot shed their pollen until the stigmas have withered, there is no possibility of self-pollination. Pollen is shed during the second night. Meanwhile, sexual satisfaction has induced other changes in the inflorescence: the downward-pointing conical papillae have withered, leaving in their place a rough and wrinkled surface, up which the insects can climb and escape. By this time the spadix is no longer emitting its foul smell so that insects are not attracted to visit the same flower a second time: but no doubt some other flower is smelling just as bad and they go off to visit that.

One of the most remarkable features of the aroids is that the spadix actually generates heat. Both how it does this and why it does this were for long a mystery, and for almost as long misunderstood. It was thought for a long while that it was the heat of the spadix which attracted the insects, but this is now known not to be the case. The function of the heating apparatus is to vaporise the malodorous compounds and to assist in their dissemination. The heat itself is produced by an increase in the rate of transpiration within the spadix. The quantity of starch consumed by the spadix – several grams – within a few hours is totally out of proportion to the quantity of smell produced. The smell itself is produced by a very few milligrams of such substances as ammonia, amines, amino-acids, skatole and indole. The smell itself is a seductive device: the heat is merely required to make it work. But the whole thing is a cruel deception, for the insects receive no food, no edible pollen, no nectar, no petals – which is perhaps why they fall into the next trap.

The sheer numbers of insects that visit these flowers is quite amazing – and shows just how well the deception works. It is on record that the sexual chamber of one *Arum maculatum* inflorescence contained some 4,000 insects, and that this figure was not untypical. In a related species, *Arum conophalloides* from Asia Minor, the insect visitors are blood-sucking flies. In one spathe alone some 600 of these blood-sucking flies were found, of which 461 were identified. These represented a mere three species of insect. 427 of the individuals were female. Thus, although the scent of most arums is not particularly noticeable to humans, it is plainly very noticeable to insects, and highly specific both to sex and genera.

Other members of the *Araceae*, the arum family, have some intriguing variations on this theme. It was among the closely related genus *Arisaema* that the phenomenon of 'window panes' was first discovered. These window panes are simply areas within the trap which appear to be transparent when viewed from the inside. Insects, once they have entered the flower, are lured still further into the trap by the light coming through the window: by the time they realize that they cannot, after all, make their escape through the window, they are ensnared with no means of escape until the flower chooses to release them.

Other members of the genus *Arisaema* use light to effect a rather different type of trap. With these there seems to be some special type of tissue which reflects and refracts light, concentrating it to a considerable power of illumination: insects lured to the spadix by its smell, are lured into the trap by this strange glow. The effect of a glowing light within the chamber is enhanced by the dark colours surrounding it. The plant in which it was first noticed, *Arisaema laminatum*, has a spathe which has a green hood and a dark purple entrance. In *Cryptocoryne* (now *Arisaema*) *griffithii* both the window pane effect and the 'inner glow' methods of luring the insects into the trap are used. Then there is the giant of the family *Amorphophallus titanun* (which being literally translated means the titanic love-tool) a huge thing with a spadix as thick as a fat man's thigh and some two metres tall. It has the usual fecal smell associated with this group, and is pollinated by large beetles. Once these have been lured into the trap they are unable to escape by an overhanging ridge low down on the spadix which has such a sharp ridge that the beetles cannot negotiate it, and simply fall back down into the trap: again and again.

Another plant in the same family has a trap which is not merely passive, but decidedly active: indeed almost animal. This is *Typhonium trilobatum* which is pollinated by tiny little beetles no more than half a millimetre long. These innocently come to investigate the spathe, gradually working their way down it towards the female flowers: once there, the spathe constricts just above them, imprisoning them. The captive beetles have to remain there for a whole day before they are allowed to escape, and even then they cannot get away without the plant having its will of them. During the time the spathe remains constricting the beetles in the prison, the male flowers shed their pollen, which collects in a dusty ring above the constricted area. When the spathe releases its hold on the beetles it does so by only opening by the tiniest amount, so that in order to escape the beetles have to make an undignified exit through this mass of pollen.

Even the one-way traffic system does not seem to have been a human invention, but appears rather to have been invented by certain flowers many millenia before man came on the scene at all. A typical one-way system occurs in the *Arisaema* relative *Colocasia antiquorum*. In this a strong fecal smell emitted by the spathe attracts insects which enter at the base of the spathe, instead of entering by the front door, as is common with other species. Having entered the base of the spathe, a constriction above prevents them from going any further. They therefore remain in the region of the female sex organs, attracted by the smell. Later in the day the smell fades but the insects are prevented from going any further by the constriction above them, and are prevented from escaping by means of the route by which they entered since once the smell fades, that too closes up behind them. During the night the constriction above the insects opens, admitting them to a second chamber. It is in this second chamber that the insects encounter the male sex organs, which shower them with pollen. On the second day, once the insects have been thoroughly dusted with pollen, they are finally released through the top of the spathe by the opening of the final constriction. In some species the flow of the insects is reversed. They enter at the top of the flower, climb down the spadix, crawling first over the male and then over the female flowers, making their exit through a hole in the bottom of the spathe. This seems a less satisfactory method of traffic control, since it induces a high probability of self-pollination.

Some of the species within *Arisaema* have even more macabre methods of ensuring pollination. *Arisaema leschenaultii* is typical of this group. The group produce both male and female flowers but on different plants: the catch is that only the males have the exit hole at the bottom. So far so food. The pollinating insects, mainly tiny fungus-gnats, go through the one-way system, and in so doing get covered in pollen. They are then lured into a female flower, which not only has no exit hole, but has the flowers very tightly packed, with the stigmas sticking right out as far as the embracing wall of the spathe. The poor fungus-gnats, force their way down deeper and deeper into the flower in search of the nonexistent exit hole, get jammed between the stigmas and die.

It would seem that the idea of using traps to ensure pollination arose more than once in the vegetable kingdom, and each time apparently quite independently. There are orchids in which the total flower structure, its method of trapping and its general appearance, are so like those of the *Arisaema* that they could easily be mistaken. Those of the milkweeds (*Asclepidaceae*) and the birthworts (*Aristolochiaceae*) operate rather differently.

Those orchids which are most like the fly-trapping *Arisaema* are removed from them by an enormous period in evolution. Yet the similarities are quite amazing. The orchid genus bearing the closest resemblance to the *Arisaema* is *Paphiopedilum*, itself the tropical equivalent of the temperate genus *Cypripedium*: confusingly, both are known as lady slipper orchids. What the two genera have in common is the same one-way method of trapping. In this group of orchids only two of the stamens are fertile, the third forming the front of the column. Of the outer whorl of stamens one is transmogrified into a thick, petal-like staminoide, which arches over the stigma. The lip forms the characteristic pouch from which these orchids get their common name. Nectar secreted inside the lip attracts visiting insects. These enter the lip through the obvious opening, but are unable to escape by the same route owing to the slipperiness of the side of the lip and the distinctive way in which it is enrolled at the edges. The insects then cast about for some alternate path of escape, and are attracted by light coming through translucent window panes in the sides of the lip. In heading for these the insects come into contact with the column, where a multitude of hairs on the floor of the lip provides them with a secure foothold. The visiting insects are then guided to their only escape route, which is really rather like an assault course. They have first of all to squeeze their way under the long, rough stigma, finally to crawl out through one or other of the two small holes on each side of the base of the overhanging staminoide petal, in the process of which they have pollen smeared onto them. The whole system is highly sophisticated. The insect visitor is carefully guided first past the stigma, so that any pollen it has collected from a previous visit will pollinate this stigma: it then collects pollen again on the way out and, since there is usually only one flower open on each stem at a time, there is virtually no possibility of a plant receiving its own pollen. Again there is a graded-grains system which allows insects too small to perform the sexual job properly to come and go quite easily, but large insects, though they can enter the flower, cannot leave it and usually perish.

Among the tropical lady slipper orchids the type of pollinator attracted is usually one of the dung-inhabiting flies or beetles. Not only do these orchids emit the appropriate fecal stench, but also the prevailing colours are apt, dull browns, lurid purples, dark reds, in various weird combinations.

Not all the trapping orchids do their trapping in such a passive way: some are actually stimulated into activity when a visitor touches a sensitive area. Thus in the tropical orchid *Masdevallia muscosa* there is a sensitive area on the triangular part of the lip. As soon as an insect settles on this it rises upwards, trapping the insect. The only possible means of escape for the insect is between the lip and the column, but in attempting to get out that way it must encounter both stigma and pollinia. The trap remains in the closed position for about thirty minutes, after which it slowly opens, releasing its captive. A very similar active trapping mechanism is found in the Australian orchid genus *Pterostylis*, though in this case the plants appear to be making a deliberate attempt to imitate the Arisaemas. Their colouring is the typical dull green or red of the *Arisaema*, marked with darker, vertical stripes, again characteristic of *Arisaema*. Further, these Australian orchids have so arranged their floral parts that the uppermost sepal looks remarkably like the spathe of an *Arisaema*. Some species even emit the requisite stench. Insect visitors are treated quite roughly. If they land on the sensitive area of the lip they are flung quite violently, upside-down and back first onto the stigma, which they pollinate on the way, so long, of course, as they are carrying pollen. Having sprung the trap the insect has only one way out: an undignified backwards exit, in the process of which it cannot but get the pollinia lodged on its thorax. Apparently these orchids have even taken the unusual step of so arranging matters that each different species is pollinated by a different insect species.

The other great family of trappers are the milkweeds (*Asclepiadaceae*), but in these the traps operate on a rather different principle. To begin with, the primary seduction mechanism is nectar and not the fetid fecal odour with which we have become rather familiar in the last few pages. Curiously, although in no way related to the orchids, some features of their sexual engineering do have quite a lot in common. The flowers are radially symmetrical (not zygomorphic as in orchids) but the stamens and the huge style function in the same way as they do in a typical orchid, being fused together to form a big, rather thick organ called a column. The pollen is usually produced, as it is in orchids, in the form of pollinia connected by bands to horny clips which attach themselves (with almost pincer-like tenacity) to the legs of visiting insects. The clip and connective bands are called a translator. In the milkweeds there are five translators in each flower, and they alternate with five anthers, each of which contains two pollinia. Each translator connects two pollinia, one from each of the two adjacent anthers. Visiting insects coming for nectar find that as they try to fly away their legs are caught in slits on the column. If the insects are strong enough to get away the clips are firmly attached to the claws of the insects. (If they are not strong enough to get away they die.) Once the insect has got away what is probably the most extraordinary part of the whole operation occurs: the clips, attached to the claws of the insects, twist themselves by spontaneous movements of the connecting bands, through a complete right angle. This is necessary for the next stage of the trapping game. At the next flower visited by the insect the pollinia wedge in the slit and break off, finishing up on the stigmatic area of the flower. In some members of the family it is to the proboscis of the visiting insect that the clips become attached.

Probably the most fascinatingly beautiful of the flowers which find bondage necessary to their sexual purposes are those of the largely succulent genus *Ceropegia*, of which there are species native to Asia, Africa and Australia. Their flowers, though generally rather too small to be observed in their full beauty by the naked eye, but readily enough observed through a hand-lens, come in delicate shades of grey, green and brown, and are fashioned into elegant tubes with delicate lobes which unite at the tips to give the effect of some fairy lantern – altogether an amazing piece of floral engineering.

In spite of their structural differences, these plants employ the same basic techniques of seducing insects to serve their purposes as do the *Arums*: indeed, the similarities between these two groups of plants far outweigh their differences of appearance. The insect is attracted to the flower by scent: although the scent is usually almost imperceptible to us, it is plainly noticed by insects, since in approaching the flowers their flightpath is typical for an insect approaching a scent source. Insects seem invariably to alight on the area which is producing the scent, and therein lies their greatest mistake. Because the scent-producing area either coincides with the slipway or is immediately adjacent to it. The insects, attracted by the dark interior of the flower (this often being darkened by a deep reddish pigment) slip and fall straight into the trap. The slipway is constructed in almost exactly the same way as in the arums, of downward pointing conical papillae coated liberally with a lubricant.

Having fallen into the trap the insects then move on into the copulatorium, which is usually wholly or partly illuminated by window panes similar to those found in the *Arisaemas*. These windows normally form a ring round the sexual organs of the plant. The insects, feeling safe, in their well-illuminated prison, happily climb the column at the centre of the windows, and quietly set about sipping the nectar. When they have drunk their fill they withdraw their heads which get caught in a grove on the column: the only way they can get their heads out of the groove is to pull hard enough to bring the pollinia clips too. If the insect already has pollinia clips on its head, its head gets caught in a groove lower down on the column, and the pollinia is pulled off on the stigmatic part of the column. After a day or two the flower relents and releases the prisoners it has held, and it does this, firstly, by gradually tilting itself from the vertical to the horizontal and, at the same time, allowing the papillae which prevented the insects from climbing back up out of the trap to wither away, leaving the insects a clear exit.

From the point of view of the flower seducing the insect this type of flower must be among the greatest masterpieces of floral engineering in the entire plant kingdom. It appeals by one means or another to almost every important insect instinct. Successively it lures the insect to it by a scent which the insect associates with mating, then by a dark cavity, an ideal place for the insect to lay its eggs, then uses light to lure it into a trap from which the insect seeks to escape and finally ensures pollination by offering the long-suffering and much-tricked insect food. All in all, quite a *tour de force*.

Certainly for those flowers which can only obtain sexual satisfaction through bondage, that bondage is as necessary as it is to humans who can find no other satisfactory *modus amore*.

29 Coprophiles and Necrophiles

DUNG AND DEATH are not subjects upon which we usually have much cause to contemplate, and in this we are in many ways luckier than our forefathers. It was not, after all, so very long ago that our forefathers were throwing the waste products of their digestive systems (usually politely called 'night soil') out of the window every morning into the city streets beneath, where those less fortunate had to wade ankle or knee deep in these deposits. Modern conveniences for disposing of these unwanted products have only become the everyday fixtures of our homes during this century, and even then only in societies which would normally be considered civilized. Even the Victorians, who considered sex an unmentionable subject, accepted death and dying as something that quite naturally took place, usually in the home at that. But today, where sex is the normal topic of conversation wherever two or three are gathered together, death is the forbidden topic. Just as we neatly flush the waste products of our bodies out of sight, so too we remove our dying to do their dying on their own, in the decent cleanliness of some well-scrubbed terminal ward. Neither death nor defecation is any longer a part of our everyday consciousness.

Apart from sex, we seldom consider our bodily functions, except when something goes wrong with one of them. Yet were we to do so we should quickly realize that what man eats he excretes; that the same is true of all other animals: and that whatever lives must die, and that what dies must be dealt with. It is only a very thin and totally artificial veneer of sophistication that hides these realities from us. In the wild these things are dealt with quickly and cleanly. Flies buzz and hover above the droppings of animals, while bugs and beetles carry off more dessicated matter, other insects and organisms gradually draw the remnants into the soil. In hot climates the dead are quickly dealt with either by those who killed them, followed up by scavengers of one sort or another (what the lion doesn't eat the hyena does, and what he doesn't eat the jackal does, and what the jackal leaves the vultures strip clean; then maggots chew their way through the sunbleached bones) or by some other natural sequence of scavengers.

If we find ourselves slightly offended by being reminded of these things, it is because we have for too long turned our faces away from these realities, and have for too long tried to forget that we are only animals like other animals. For in the natural world almost every possible ecological niche has been filled. And among those ecological niches are dung and death. And if the creatures that deal with these things are small, highly mobile insects, the chances are that some flower will try to exploit their natural preferences in order to obtain pollination.

The creatures that wallow in dung are called coprophiles (from the Greek *Kopros* = feces + *philein* = to love), although some of those which are attracted by flowers are more properly copro-lagniacs: after all, if a flower wants to seduce something to perform its sexual functions for it by deceiving the insect into thinking it is dung, it might just as well go all the way and give the victim it lures some satisfaction on the job.

The most common means by which flowers attract coprophilous or coprolagniac insects is by emitting a strong fecal smell. This is usually combined with some other device to ensure that the visiting insect, having been seduced by the repulsive smell, does not promptly escape once it realizes the deception. The commonest occurrence of the fecal smell as a seductive device is among those flowers which trap their insect visitors and is dealt with here.

By far the most curious group of plants which use and rely upon the coprolagniac instincts of some insects are the stapeliads (*Stapelia* spp), which are somewhat cactus-like succulent plants, natives of Africa and southern Asia. In these curious plants the flowers have done just about everything evolutionary possible to make themselves resemble rotting meat, or stinking droppings, not only in terms of smell but also of colouring, patterning and even texture. This has given rise to some quite remarkably repulsive flowers: imagine a flower 40 cm (16 inches) across that not only looks like rotting meat and smells like rotting meat, but attracts flies as copiously as rotting meat and even has the eggs of the flies laid on its textured petals, and the larvae of the coprophilous flies crawling over it. Indeed, so powerful is the impression of rotting flesh that among some of the native peoples of Africa legend has it that these fecal flowers are pollinated by visiting vultures. Unfortunately insufficient research has been done on this group of flowers positively to deny this interesting legend. In other species the resemblance, just as carefully created, is of animal droppings, and these are often pollinated by scarabs and other dung-beetles. It would seem that the stench emitted by the stapeliads, including the related genera *Huernari* and *Caralluma* are essentially similar to the breeding scents of the insects which they attract. Since the eggs of the insects are most usually laid near the centre of the flower (and can often be readily seen with the naked eye), it seems probable that the visitors come for this purpose as much as for any other, and that while some of the pollinia clips must get attached to the legs of the egg-laying insects, some may also get attached to the larvae as they crawl over the flower.

If, as humans, we find something rather repulsive in the idea of flowers which have their sexual organs right at the centre of a flower which is a deliberate imitation of a foul-smelling animal dropping, that is because we have a poor understanding of ourselves. After all, we in common with most other animals, copulate by means of an organ which in the female is placed midway between the organ through which she micturates and the organ through which she defecates.

30
Advanced Techniques

IT IS A CURIOUS THING that though most of us have, at least in our childhood, collected lamb's tails and pussy willow, and though most of us have probably noticed the deposits of yellow dust that these have left upon precious polished surfaces, few of us have realized that that yellow dust is pollen, the precious sperms as it were of the flowers we have so carelessly plucked. Fewer still have realized that that yellow dust represents one of the most highly specialized sexual techniques of the plant kingdom or that, in some species, it points the way to the future.

Wind pollination in conifers is regarded as primitive: they adopted wind pollination in order to avoid having their private parts eaten by the beetles which were the dominant insects of the era during which the conifers evolved. But the conifers are by no means the only plants in which wind pollination occurs, though they are the only group of plants in which it is a definitely primitive trait.

What is curious is that wind pollination should crop up again in higher plants, among both the dicots and the even more modern monocots, yet in neither group is it regarded as a retrogressive step. Rather it is a matter of completely different plants reaching very similar conclusions by completely different paths. In the conifers wind pollination may well have evolved because there was no other reliable way of transferring sperm to egg: the cone, after all, was a considerable elaboration on the organs that went before. In those flowering plants that are wind pollinated, the seductive aspects of showy petals and so on have been deliberately discarded to achieve a streamlining which makes wind pollination a more certain way of maintaining the species.

The long yellow catkins, which are the male flowers of the hazel (*Coryllus avellana*) are one of the earliest harbingers of spring, and it is for this reason that they are so eagerly gathered. Yet few people, picking these catkins, realize what it is they are actually picking. What they are actually picking are the male sexual organs

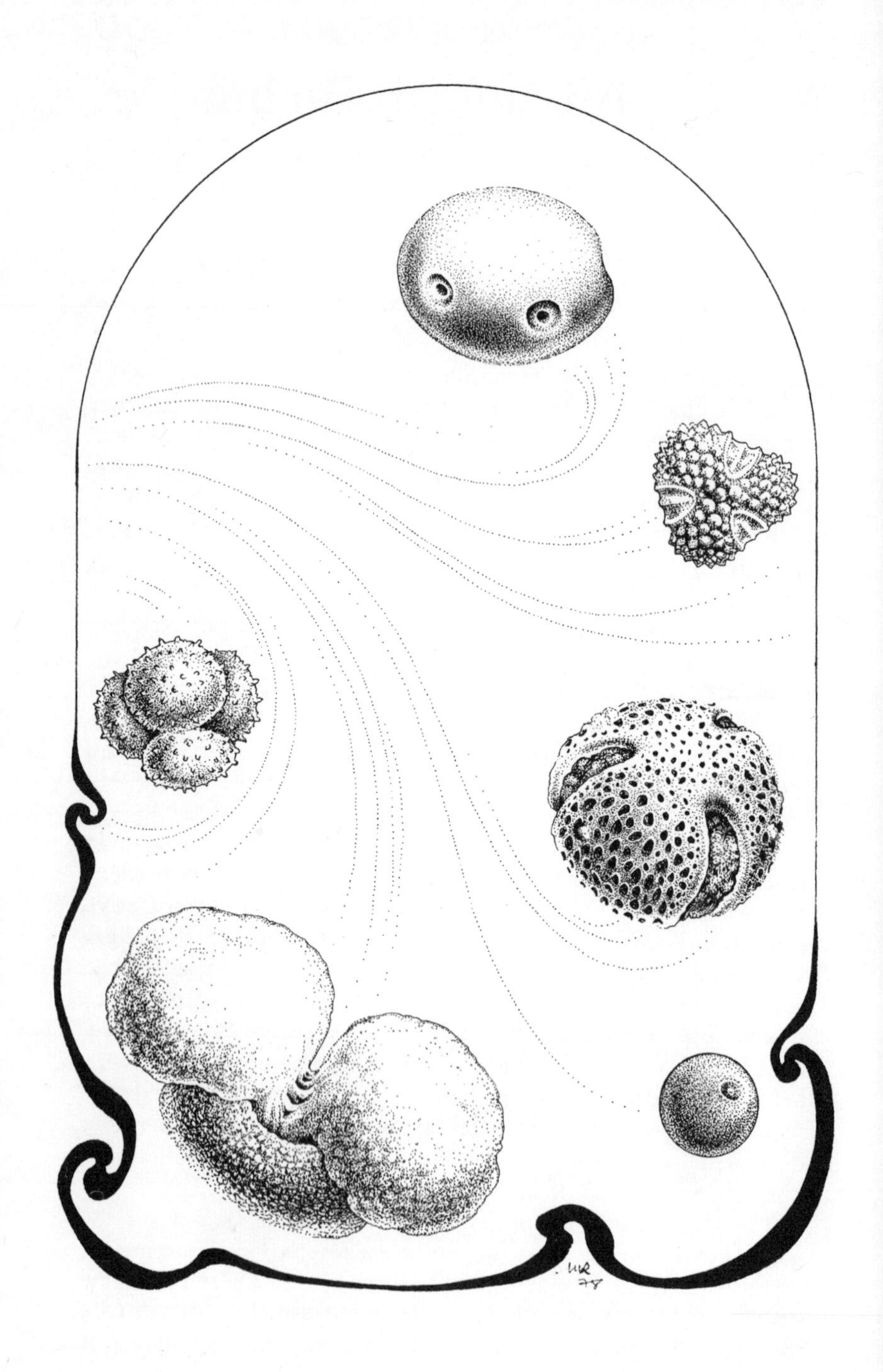

of the hazel, each catkin consisting of numerous male flowers, each flower of which is reduced to its barest essentials. The flowers have no recognizable floral parts, but consist of little more than stamens grouped in pairs and protected between scales, strung out on a loosely hanging tassel that may be shaken by the wind. In still air, very little pollen falls off the catkins: the male organs need the stimulus of moving air joggling the tassels to release their precious semen, that yellow dust that clings so tenaciously even to a cleanly polished or glass surface.

While the male catkins of the hazel are, to human eyes, relatively noticeable, the female catkins usually pass unnoticed, yet they are worth looking for. The female flowers are also produced in what is technically a catkin, but a very much reduced one, scarcely noticeable as a catkin at all. Indeed, if you do notice them, all you will see with the naked eye are the brilliant scarlet stigmas protruding from the tips of catkins so reduced as to appear to be no more than a swollen bud. Once you have spotted them, these female flowers are quite showy, and fairly easy to pick out on a sunny day. If they are fertilized they develop in time into a one-seeded nut surrounded by a coarse leafy structure, the familiar hazel nuts.

Male and female flowers are borne together on the same tree in the hazel, though they are produced on different parts of the twigs. In the sallows or pussy willow which flower at much the same time, male and female flowers are produced on different plants, and this is true of all the willows, including the familiar weeping willows. Both male and female flowers consist of catkins, but the catkins differ greatly according to sex. In the male the catkins are stiff and usually erect, up to 6cm long and about 1 cm across. The individual flowers are produced in the scales of the catkin, one flower to each scale, and the flower itself is reduced to its barest essentials, little more than two stamens with a nectary at the base. In the female the flowers consist of little more than an ovary with a short style and two stigmas. Because of the nectaries, the catkins are visited by insects which play an important part in the pollination of willows, though it would seem that the greater part of the pollen is dispersed by the wind. In the poplars, which are closely related, except that their catkins usually hang downwards, pollination is accomplished entirely by the wind.

In the birches (*Betula*) the male catkins are formed in the early winter, but do not mature until spring. The male catkins are short and firm through the winter, but gradually expand in spring, releasing their pollen as they do so. Typically the flowers are borne in threes, each flower consisting of one pair of deeply divided stamens, with a small, bract-like perianth. The female catkins are not borne at the tips of the twigs like the male catkins, but on short, spur-twigs. Unlike the male catkins, which dangle loosely in the wind, the female catkins remain erect and rather tightly formed. The female flowers occur in the axils of the catkin scales, each flower consisting of merely an ovary bearing two styles. As in all wind-pollinated plants, male flowers vastly outnumber female flowers. And the quantities of pollen produced are absolutely astronomical. It has been calculated that a single male silver birch catkin will produce something in the region of 1,300,000,000 grains of pollen. If life were long enough it would be interesting to count the number of male catkins on a mature tree. Because the total production from one tree must be millions of millions of grains.

It is curious that in some plants we notice the catkins, but generally do not particularly associate them with the fruits, while in other trees we are all well aware of the fruits, but fail to notice the flowers. We all know an acorn when we see one: most of us know an acorn even when we don't see it. But can you recall what the flowers of an oak tree are like? Or those of a beech tree? Yet both of these trees bear catkins in the spring, before there are leaves to obscure them. The male catkins of the oaks are actually rather more conspicuous than those of several other catkin-bearing trees. The catkin scales are very small, and the individual flowers have rather larger perianth segments than is usual in catkins: they also have about six stamens in each flower, which is more than most catkin-bearing trees produce. Perhaps the reason we overlook them is simply that they are almost exactly the same colour as the new leaves which are just beginning to burst from their buds at the same time that the catkins are produced. The female flowers are borne in a short, spike-like catkin, and each flower is surrounded by a large scale which in time becomes the cup of the acorn.

In the common European beech the catkins are long-stalked, tassel-like contraptions, and again the reason we overlook them is that they are the same colour as the new leaves, and are produced at exactly the same time. The female flowers are produced in pairs, each flower being surrounded by a scale which, in time, grows to form the four-valved cupule in which the nut is held.

Two of the earliest wind-pollinated trees to produce their catkins are the alder (*Alnus*) which is in all essentials very similar to the birch, and the elm (*Ulmus*) which is in just about all respects quite unlike any other wind-pollinated tree mentioned so far. For a start, the individual flowers contain both male and female sex organs. However, the flowers use a time-lag method of contraception, so that there is little chance of a flower being fertilized from its own pollen. Each flower consists of a bell-like perianth segment divided into either four or five lobes, containing either four or five stamens, and a one-celled ovary with two styles. The stamens very often stick out so far from the rest of the flower that one could easily mistake the flowers for being completely male.

Probably the most interesting of all the wind-pollinated trees with which most of us are familiar is the common ash. Why it is interesting is that almost all the other members of the family to which it belongs, the olive family (*Oleaceae*), are insect-pollinated. Some of them are well-known; lilac, jasmine, privet. But by comparison the flowers of the ash could easily pass unnoticed. The corolla is lacking, but there are two stamens and a long, narrow ovary which bears two rather large, black-tipped stigmas. Some trees bear only male flowers, others only female flowers, and some flowers containing both sexes; to make matters more complicated, the tree may vary in its sexuality from year to year. The manna ash (*Fraxinus ornatus*) in contrast with the common ash, still retains its corolla, and the scent so typical of its family, and is still pollinated by insects. The particular interest of the common ash lies in that it is a plant that plainly evolved a mode of sexuality dependent upon insects, and has only recently abandoned it in favour of wind-pollination. It is the loss of the corolla that is so conspicuous since it seems to occur in almost all the trees which have abandoned insect pollination in favour of wind-pollination.

The trees, however, have no exclusive rights in the matter of wind-pollination. It is also surprisingly widespread among herbs (as opposed to woody plants), and is particularly common in aquatic and waterside plants. What is most striking about the herbs that are wind-pollinated is that they belong to such widely differing plant groups. Plants as diverse as the common stinging nettle (*Urtica dioica*) on the one hand and its natural sting-remedy, the almost-as-common dock (*Rumex*) both being wind-pollinated, as are the rock roses (*Helianthemum sp.*) ling (*Calluna vulgaris*) and the mistletoe (*Viscum album*). For some reason there is a particularly large proportion of waterside and aquatic plants which are wind-pollinated, though as yet it is unclear why this group of plants should have developed those characteristics which are associated with pollination by the wind, especially as the plants come from widely differing families. There is dog's mercury (*Mercurialis perrenis*), which is a member of the spurge family (*Euphorbiaceae*), and the mugwort and some of its relatives, which are all members of the *Compositae*. Salad burnet (*Poterium sanguisorba*) belongs to the rose family (*Rosaceae*), while the meadow rues (*Thalictrum* spp.) belong to the buttercup family (*Ranunculaceae*). The cattails or reed-maces belong to the *Typhaceae*, and so are relatively close to the grasses. In evolutionary terms there is no single ancestor from whom all these plants could have derived their wind-pollination habit. The conclusion is that it must have arisen independently in each of these families.

There are, however, two very large groups of plants in which wind-pollination is the norm, and it would seem that these two groups must have set out deliberately to exploit this mode of pollination. There can also be no doubt that for these two groups of plants, wind-pollination was the right choice, probably thereby taking advantage of a particular evolutionary niche which till their coming had never been fully exploited. These two groups are the grasses and the sedges, and there can be no doubt that the grasses at least, are among the most successful of all plant groups. They are a mere 60,000,000 years old (compared with the conifers which are about 300,000,000 years old) and yet they have adapted themselves to colonize almost every part of the earth's surface: certainly they could never have achieved that by any other means than wind-pollination. Try to imagine a world without the grasses, a landscape without grasses, a roadside without grasses.

A lot of people tend to link the grasses and the sedges together in their minds, but this is probably only because there is some superficial similarity in the flowers of both groups. In fact the grasses belong to one family, the *Poales* (still sometimes called *Graminales*), while the sedges belong to a different family, the *Cyperales*. The only thing they have in common is that way, way back in their ancestry both families evolved from the *Liliales*: but that does not explain why both families are wind-pollinated; after all, the orchids and the iris also evolved from the *Liliales*. The

grasses appear to have evolved directly from the *Restionales*, in which there are clear signs of a reduction of the floral parts of the type which appear to have been taken still further in the grasses. The sedges seem to have evolved directly from the *Juncales*, the family which includes the rushes. Each evolved similar pollination mechanisms completely independently of the other.

So specialized are the flowers of grasses that they have a whole specialized vocabulary to themselves. The flowers are produced not in an inflorescence, but in spikelets, each of which is enclosed at the base by a pair of chaffy glumes. The individual flowers in the spikelet alternate on either side of a central stem called a rachilla. Each flower is enclosed in glume-like things called lemmas and paleas, while the flower itself is reduced to the barest necessities: merely a single ovary with a pair of feathery stigmas and three stamens with slender filaments and large, rather loosely hinged anthers plus a pair of tiny, scale-like things called lodicules which swell and in so doing cause the flower to open. These lodicules are probably all that is left of the showy floral parts of the distant lily-like ancestors of the grasses.

The sedges are superficially similar, but there are some important differences in the flowers. The flowers of the sedges are borne singly on the axils of the scale-like glumes which together compose a sort of catkin. In most of the sedge genera the flowers produce both male and female organs, and there are usually three stamens, a single ovary containing a single ovule, with a style divided into three long, rough stigmas. However, in the widespread temperate genus *Carex*, male and female flowers are borne separately, as they are on some of the catkin-bearing trees, and the catkins have a very similar appearance.

Among the grasses the use of a contraceptive clock to prevent fertilization of a plant by its own pollen is taken to its most highly developed stage. A grass does not simply open its flowers, shed its pollen, and then give the stigmas a turn. Nothing so simple. Each grass is highly specific as to the time of day when it will or will not open its flowers to shed pollen. Many open in very early morning, or at dusk, since these are the times of day when, due to the changes occurring in the air temperature, convection currents are set up which ensure that there will be wind to disperse the pollen. But they are even more specific in their flowering habits than that. One species may, for example, open its flowers to shed pollen at precisely 6.35 am, continue shedding pollen for thirty-seven minutes, then close again till the next day. Some may open their flowers briefly twice or even three times during a day. There really is no other group of plants which seems to be able to show anything comparable in the degree and sensitivity of their behaviour.

It has been said of the grasses that, because of their highly specialized way of life, including their compulsive punctuality concerning sexual matters, they are the plants of the future, the plants that will inherit the earth. Which may be true because the grasses, coming late on the scene, into a world in which insect-pollination was already well established, had the option between evolving in the direction of more and more specialized insect-pollination, as have their rival and probably coeval group, the orchids, or throwing their pollen and their seed to the wind. In choosing to abandon themselves to the wind they had to sacrifice their showy floral parts, and they had to develop a lighter pollen than other plants, yet these sacrifices seem to have been worthwhile. For any plant that is pollinated by insects can only spread and endure within the range within which those insects live. This is a severe limitation, both in broad climatic terms, and also in microclimatic terms (few insects, for example, penetrate the darkest depths of forests). But the wind blows everywhere, to a greater or lesser degree. And pollen has been found as high up in the atmosphere as 2,000 km, and over 5,000 km from its nearest possible source plant. Perhaps pollen that can travel quite those distances has gone too far, but it does serve to illustrate the point.

31
Sub-Aqua Sex

MOST PEOPLE HAVE HEARD about the birds and the bees. And many people know that other creatures such as moths and even mammals often get in on the sex act of flowers. Not quite so many seem to know about the pollination not only of conifers but also of the higher flowering plants by the wind. And very few people indeed even seem to think of water as a pollinating agent.

Which is strange, because in so many ways water seems the most obvious thing of all to use to aid in pollination. For one thing, there is more water in the world than anything else. And everything that lives is mainly water. You and I, fat or thin, are seventy-four per cent water. Besides, all life began in the water, and it is really that more than any other consideration that makes water seem the most natural of pollinating agents.

Yet curiously enough water is probably the least exploited of all the available pollinating media. As Agnes Arber wrote in her classic book *Water Plants* (1920): 'The most notable characteristic of the flowers of the majority of aquatic angiosperms is that they make singularly little concession to the aquatic medium . . .' And how right she is. For the great majority of flowering plants that live out their lives in the water, are insect-pollinated or wind-pollinated. Only a tiny number of those plants that live in the water actually use the water to pollinate them, and they are the exceptions. Yet those modifications that would be needed to turn a flower into one suitable for water pollination are small and few.

There is, however, quite a large number of aquatic plants which use the surface of the water as a medium for pollination, and this seems to be almost a compromise between wind-pollination and water-pollination. A typical example of this sort of compromise situation occurs in the ribbon weed (*Vallisneria spiralis*). In this plant male and female flowers are borne on separate plants. The flowers of the male are reduced to two stamens tightly embraced by three sepals. These are tiny, but are borne together in a tubular spathe produced well under the water. At maturity, the male flowers break loose and float to the surface of the water, where they open; they are then swirled along by water currents or by the wind. The female flowers, on the other hand, remain firmly attached to the parent plant, as needs they must. Each female flower is borne to the surface of the water on a long, slender peduncle. The flower consists of three thick sepals and three thick, sticky and totally unwettable stigmas, with a very reduced spathe. The whole structure is designed so that it does not so much float on the surface of the water, as actually within the layer of molecules which form the 'skin' of the water, and are so shaped that they form a small dimple on the surface of the water. Any male flower that happens to drift close enough to a female flower, slides down the dimple in the surface of the water and finishes up in the arms of the female flower, its stamens firmly stuck to her stigmas.

In the Canadian water-weed (*Elodea canadensis*) there is a small refinement on this system. Male and female flowers are once again borne separately, but both are borne on the surface film of the water. The male flowers are each made up of six water-proof perianth segments, and nine stamens. The flower opens abruptly, the water-proof perianth segments lifting the nine stamens well clear of the water. On exposure to air, these dehisce, and then suddenly explode, showering the water with pollen grains. The pollen itself is wettable, and deteriorates very rapidly when it is wetted, so the grains are borne above the water by a thin layer of air trapped by densely set minute spines. The pollen grains drift around for a while, and if they are lucky, happen to drift close enough to a stigma from a female flower to slip down the depression in the surface tension of the water, they will end up on the stigma. All of which, though very ingenious, seems to be a very chancy way of getting yourself pollinated: a point of view with which the plants seem to concur, since male plants are exceedingly uncommon within the species, and the plant normally reproduces vegetatively.

These plants are using only the surface tension of the water to aid them in their sexual endeavours and anything, like a fallen leaf, that happens to float on the surface of the water, will be moved as much by the wind as by the water. Indeed, no adaptations have taken place in these mating mechanisms of these plants that would not qualify them just as well for wind-pollination. True submerged mating is still a rarity – and probably a novelty.

One group of plants among which it seems to be well established

are the water starworts (*Callitriche*). The great majority of species are wind-pollinated, and those in which true underwater sex takes place have now been moved to the sub-genus *Pseudocallitriche*. In the terrestrial species flowering and fruiting occur in the upper axils; in some species which may be regarded as intermediaries, while the plants live mainly in the water, flowering and fruiting occur only in those axils which are above the water. Even those species which flower and fruit underwater seem to have made no apparent concession to their watery medium. They are pollinated exactly as they would be if they were above the water, except that the water carries the grains around instead of the wind. The only concession they seem to have made to the water is the presence of oil globules in the pollen grains; these presumably play some part in preventing the pollen from being wetted. A tiny concession, but apparently all that is needed.

The horned pondweed (*Zannichellia*) seems to have made greater concessions to its watery home, possibly because it has lived in the water for longer than the other plants mentioned. The flowers are very specialized in both form and function. The specialization seems to be merely a matter of reduction, in this case, almost to absurdity. The male stamen releases pollen grains into the water, and these sink slowly downwards, coming to rest on the stigmas, where they slide down the stigmatic canal to effect pollination. There is however, one added touch of sophistication to this simple performance.The pollen grains nearly always start to germinate before release from the male organs. By the time they are released each pollen grain has already started to grow a pollen tube (which is the proper equivalent of the penis in the male of our species). It is this appendage which hooks onto the stigma.

Among the eel-grasses (*Zostera*) there is an even higher degree of specialization. Typically the eel-grasses grow along coasts in the zone between high and low tide, and equally typically, they produce long grass-like leaves from rhizomes that creep through the mud or sand. The interesting thing in the sex lives of these plants is that the pollen grains, which are released into the water flowing over the eel-grass when it is submerged, have adapted themselves to their environment to a high degree: they are not rounded grains like other pollen grains at all, but are thread-like, and of the same density as the sea-water in which the parent plants live. Once one of these pollen grains, drifting in the tide, encounters any narrow object which might be a stigma, it reacts positively and quickly, curling itself round it. It is in this way that the pollen grains ensure pollination, and also ensure that the sea tides, which so kindly brought male and female elements together, do not brutally force them apart again.

In general it seems that the first steps towards pollination by water are much the same as those for pollination by wind, notably the reduction of the floral parts of the sex organs. More particularly it seems that a radical adjustment of the pollen grain itself is necessary for submerged sex to work out properly.

The evidence suggests that those plants which have made the adaptation to an underwater sex life have done so in the relatively recent past. While, on the one hand, it would seem that water is under-exploited as a pollinating agent, it is worth remembering that the amount of water that is sufficiently shallow for plants to colonize is relatively small. Sub-aqua sex may be an interesting divergence from the norms of sexuality in plants, but it seems likely that it will remain, at least for the foreseeable future, merely a minority interest.

32
Masturbation

THE WORD 'MASTURBATION' always sounds much more like a moral judgement than a mental or physical activity: there is something positively pejorative about the word. The moral overtones are merely historical. In fact, the meaning of the word, as derived from its etymology, is to defile or abuse oneself. There are of course numerous ways in which one can defile or abuse oneself other than those methods of sexual self-satisfaction for which this term is used.

Whatever any word actually means in the strictest dictionary terms, it nearly always carries connotations of some kind, good, bad, beautiful or whatever. Compared with the rate at which a word like 'sentimental' has changed its connotations, the moral attitudes to masturbation have changed with phenomenal speed. The word seems, perhaps not surprisingly, to have had a meteoric rise to popularity during Victorian times. The Victorians, of course, took the sternest moral view of masturbation. The Good Lord in His Wisdom had provided man with sex so that he could multiply: which he has done with rather too much success since Victorian times. Any other use of sex was ethically abhorrent. The only thing more wicked than using sex for pleasure not procreation, was the almost unthinkable thought of using it to please oneself. Indeed, in such dread was this idea held that young ladies of good breeding were not even allowed to wash between their legs for fear lest they feel some pleasure at the sensations this might induce. By the turn of the century, thanks to the frequent admonitions of headmasters and housemasters, it was part of the popular folklore of every public school that if you were to masturbate 'it would fall off!' By the end of the 1914/18 war, almost every Tommy knew that the consequences of masturbation were not in fact moral but physical: blindness and deafness were the least of the physical disabilities it could induce. By the end of the 1939/45 war attitudes had changed again: blindness and deafness were unlikely; it was more probable that imbecility would result. In the years that have passed since then attitudes have changed still further. Not only is there nothing wrong with masturbation in this day and age, but progressive-thinking psychologists and others believe that its practice should be actually encouraged; one should masturbate as frequently as possible, the thinking being that if practice makes perfect, the more one masturbates the greater one's ability to enjoy any sort of sexual experience will be: like an athlete forever in training for the competitive event. Of course should it turn out that the earlier thinking was right, then there is a whole generation of the sexually liberated who will no doubt grow up blind, deaf, dumb and probably permanently twitching from nervous exhaustion.

It is devoutly to be hoped that flowers do not suffer the same moral malthinking as mankind, for in a large number of flowers something akin to masturbation is not uncommon. That is, provided that one takes masturbation to mean self-defilement, and as something traditionally only used as a release from tensions by those deprived of the companionship of the opposite sex – a last resort. Where it differs from anything in which mankind can indulge himself is that, whereas in man in common with most other animals, it is normal to have only male or female sex organs on any one individual, among the flowering plants it is the norm for each flower to have both male and female sexual organs. The consequence of self-indulgence could therefore be conception, a possibility which mankind does not have to contemplate when he or she masturbates. It is therefore normally only in desperation that flowers are, in some cases, literally reduced to helping themselves.

Because most flowers contain both male and female sex organs, there are elaborate mechanisms by which they try to prevent self-pollination from occurring. The most common method is quite simply for either the male or the female organs within any particular flower to ripen one before the other. However, there is a problem: while animals can move around and meet to mate another day, a flower is an ephemeral, passing thing whose sole reason for existing is to get itself pollinated, thus ensuring sufficient seed for the continuation of the species. If no insect visitors come, is the flower simply to die unrequited, unsatisfied?

The answer is both yes and no with an enormous number of maybes in between. There is a small number of plants in which self-pollination is actually and absolutely a physical impossibility. And there is also a small number of plants in which self-pollination is the norm. The great majority of flowers fall between these extremes.

In plants self-pollination is not necessarily the same as self-fertilization, any more than that copulation in humans invariably results in conception. Botanists distinguish between self-pollination (autogamy) and self-fertilization. A far larger number of plants can pollinate themselves than can fertilize themselves. In many flowers whose arrangement of male and female organs is such that self-pollination can occur quite easily, a self-incompatibility mechanism prevents matters from going any further. The way in which this usually works is that, for fertilization to take place, the pollen once deposited on the stigma grows a pollen tube which penetrates the style and fertilizes the ovum: self-fertilization is prevented by the production in the stigma of hormones which inhibit growth in pollen from the same plant.

In general flowers only resort to self-fertilization if all else fails; in many orchids for example, the male and female sex organs are well separated so that there is little possibility of self-pollination occurring: this is particularly true of some of those orchids in which the stamens act like triggers, flicking pollen over visiting insects. If no insect visitor comes before the flower dies, in the act of casting the corolla it will itself trigger the stamens so that they actually touch the stigma. A similar mechanism operates in the cup-and-saucer vine, *Cobea scandens*.

Some plants seem equally happy receiving their sex either wav. Some of the legumes, in particular the garden pea (*Pisum sativum*) and the French or snap bean (*Phaeseolus vulgaris*), though highly adapted to insect pollination, are just as content to settle for self-pollination. Indeed the majority of plants grown in garden cultivation normally do pollinate themselves, and what is more produce a virtually full set of seed by so doing.

The great advantage of self-pollination is that it can be relied upon to produce with certainty a full set of seed. Its great disadvantage is that there is no exchange of genes. This is one of the reasons why self-fertilization is used only as a last resort. In any population it will usually be found that there is a balance between those plants which are outward-breeding and those which are self-seeding. The balance is maintained by the usual natural selection of pressures and processes. One point here being that, if a flower in desperation has to fertilize itself, its offspring may be luckier and get themselves cross-pollinated.

There are a number of plants in which both cross-pollination and self-pollination occur as the norm. The European wood sorrel (*Oxalis acetosella*) is a classic example, though what happens there is perhaps more clearly seen in the sweet violet (*Viola odorata*). Typically the flowers of these plants are wide-open and showy, and pollinated by bees. In spite of being self-compatible, structural arrangements normally prevent self-fertilization. However, in late spring and early summer, the very same plants produce small flowers which never open. It would seem that what stimulates them into doing this is that the increasing leaf canopy above them casts an increasing amount of shade: since plants are highly sensitive to the amount of light reaching them, it is probable that this decrease in light level triggers the release of hormones (auxins) which initiate the alteration in the flower type. What is important about this is that the same increase in the shade cast by the leaves overhead is discernible by the bees which pollinate these violets, and once the shade reaches a certain degree of darkness they no longer come into it; which means they no longer visit the violets: which is why the violets resort to pollinating themselves. Precisely the same mechanism, to achieve the same end is initiated by the same shade increase having the same effect of the pollinators in the flowers of the wood sorrel.

In those violet flowers which never open both style and stamens fail to develop properly. It is as though their development had been suddenly arrested (which probably is exactly what happens) so that instead of stamens and style springing apart when the flower opens they remain in contact in the closed flower and are still in that position when the stamens shed their pollen onto the style.

Plainly, the legitimate and illegitimate fertilization which occurs in these flowers is a response to changing conditions: so long as bees are around, legitimate offspring may be expected to be produced. Once there are no longer bees around, illegitimate offspring have of necessity to be produced. It may well be that it is the lack of a suitable pollinator that has forced certain flowers into habitual self-fertilization. Thus, for example, while the bee orchid (*Ophrys apifera*) is normally pollinated by bees in continental Europe, and uses self-fertilization only as a last resort, in Britain, where it has lost its pollinator, it is invariably self-fertilizing.

There is one very distinct group of plants in which self-pollination and self-fertilization are very common and in which they cannot be explained as having arisen as a survival response to lack of pollinators. These plants are typically annuals, typically small annuals at that, and further are typically the small annuals of unstable habitats. An extraordinarily high number of these are crucifers.

That such a large group of plants should have adopted self-fertilization as their sexual norm seems to cut right across Darwin's basically correct statement that 'Nature . . . abhors perpetual self-fertilization.' However, special sexual habits are often the result of special living conditions, and this would seem to be the case here. For one thing, these annuals tend to grow in small, isolated and very precisely defined habitats: different populations in different places may have to contend with differing limitations within their habitat. Now a true-breeding line in which every member of the population is perfectly adapted to the situation has far more chance of maintaining its hold in that habitat than a plant which shows great variety between the different individuals in the population. Further, as one habitat for a small annual gets overgrown by larger and more persistent plants, so other habitats open up. Plants which are self-fertilizing usually produce a superabundance of seed, and such seed is at a great advantage in long distance dispersal since a single plant can readily and rapidly give rise to a whole new population. And yet there is always the possibility that seed from several pure-bred lines will arrive at the same newly-revealed habitat at more or less the same time, and that at least some degree of crossing will occur between these plants before they stabilize into a self-fertilizing pure-breeding line of their own. Even such a small degree of out-breeding may well be sufficient for these particular plants to adapt to changes in climate and habitat. After all, such plants have a generation length one hundred times shorter than that of a forest tree.

It may be reassuring to realize that, among plants at least, there is as yet no evidence to suggest that masturbation produces deafness, blindness or imbecility.

33
Degeneration

THERE IS A RELATIVELY small number of plants which seem to have lost, either wholly or partially, the ability to reproduce sexually. This is normally regarded as a degenerate step, since the plants could only have reached wide distribution and survived into the present era if in the past they had reproduced sexually. What sexual reproduction secures, whether for plants or animals, is that mix of genetic material that ensures sufficient variability to enable the species, acting under the pressures of natural selection, to survive changes in climate and to some degree also in habitat.

Plants which no longer reproduce sexually have lost the benefit of this genetic mix. Whatever other methods of reproduction they have adopted instead, the offspring will be identical with the parent, and plainly this vastly reduces the chances of the species to adapt under the pressures of natural selection. Since all the plants within the species are the same, none is better adapted than any other. On the face of it, the loss of the ability to reproduce sexually would seem to have led these plants into a blind alley, for in spite of the loss of interchangeability of genetic material, some of them have adopted other means of securing some sort of genetic variation.

If sex is the normal mode of reproduction for plants, and it certainly seems to be the favoured mode, then any departure from that is degenerate. Those plants which have abandoned normal sexual reproduction are known as apomictic. The term apomixis embraces all methods of reproduction used by plants as alternatives to sex. But with plants, as with people, there are varying degrees of sexual degeneracy.

The dandelions represent one of the lesser degrees of sexual degeneracy. Curiously enough, the flowers have elaborate mechanisms not only for attracting pollinating visitors, but also well developed mechanical contraceptive devices. They have a plentiful supply of very sweet nectar, as well as the back-up of a nutritious fatty pollen, both of which are much enjoyed by bees.

The stamens appear first on the flower, and only later does the female organ come thrusting through to spread its stigmas well above the stamens, a device which not only prevents self-pollination but also seems to be designed to ensure a naturally high degree of cross-pollination. Thus far the dandelions appear very much like any other successful group of sexually reproductive plants. Yet in fact they and the related hawkweeds as well as the familiar lady's mantle (*Alchemilla vulagaris*), normally reproduce without resort to sex. The embryo quite simply starts to develop apparently spontaneously. A curious point here is that these plants do occasionally – very occasionally, almost by chance, reproduce sexually, but it is seldom. So what one is faced with is a plant, in the case of the dandelion, which has an elaborate mechanism to attract pollinators as well as a well-developed system for ensuring cross-pollination, that normally does not bother to use them. One of the fundamental laws of nature is that organs which are not used degenerate, becoming gradually more and more useless until in the end only vestiges are left of them. That this has not happened in dandelions and hawkweeds suggests that their degeneracy is of very recent origin, and that the organs have quite simply not had time to reduce themselves to vestigial proportions.

Plants such as the brambles (*Rubus fruticosus* agg.) and the cinquefoils (*Potentilla verna* and *P. argentea*) have degenerated to a similar extent, but in a different way. Their flowers are still sufficiently organized for at least a proportion of their offspring to be produced as the result of sexual activity: this gives rise to new genotypes which then reproduce non-sexually, giving rise to yet another genetically stable line.

What happens with all the plants mentioned so far is the creation of a botanist's nightmare. Because instead of having a limited number of clearly defined species, what you have is a vast number of minutely distinguished micro-species, each of which differs from the others only in the tiniest details. Indeed a full classification of some of these groups may never be reached. Generally we tend to lump them all together as an aggregate – hence *Taraxacum officinalis* agg., for the dandelion.

All that seems to be more of a nightmare for the botanists than it does for the plants. After all; all that the plants are doing most of the time is producing embryos which turn into seeds without mating. In most cases it would seem that some degree of sexual stimulation is needed to initiate the process. In many of these plants pollination is necessary, even though fertilization is not. It would seem that the growth of a pollen tube down a style, even if it never reaches the embryo, initiates hormone changes which lead to the embryo starting into growth.

The ultimate absurdity in sexual degeneration is the production by plants of fully-formed fruits in which the seeds have failed to develop at all: rather like giving birth to a still-born child without even having had the pleasure of conceiving it sexually. The bananas which we eat are a very typical example of this. In these the seeds are quite simply not formed at all. If they were we should not enjoy eating bananas as much as we do since the seeds of those bananas which lead normal sex lives are not only large but very hard.

A plant that does not produce seed at all would appear to be very close to the end of its natural span. However, the urge to perpetuate itself remains, and so sexually useless plants have to resort to other means of reproduction: edible bananas reproduce by means of plentiful suckers, while many lesser herbs such as the celandine (*Ranunculus ficaria*) and the garlics (*Allium*) reproduce by means of bulbils. In other plants, several of the grasses for example, that form of reproduction known as vivipary is widely used. The common house plant known as the spider plant (*Chlorophyllum plumosum*) also typically reproduces viviparously, producing long wands which one would expect to produce flowers but which end up by producing tiny plantlets instead, the weight of these bending the long stems down till they reach the ground, where the new plantlets take root and in time assume independence.

Other groups of plants which do not normally reproduce sexually have found other ways of ensuring survival potential by ensuring variation through sporting or mutating, which is another way of radically altering the genetic make-up of the plant, and then fixing it, at least for a time.

While undoubtedly in some cases degeneracy is a symptom of evolutionary senility, in many cases it is patently not, and has arisen through other causes. Thus among the apomictic whitebeams (*Sorbus* app.) the origin of sexual impotence is clearly a result of the hybrid origin of these plants. Possibly in some cases degeneracy has arisen as the result of the lack of a suitable pollinator, but that would not explain the large number of plants which plainly are not at the end of their evolutionary line, for which potential pollinators do exist and even visit the flowers. It seems more likely that such plants developed their non-sexual methods of reproduction during a period of extreme and difficult growing conditions – such as those which obtained during the Pleistocene era. Anyone who has seen brambles colonizing unstable ground can have little doubt that any plant that shows such exuberance is very well adapted to surviving, in spite of its sexual problems.

Maybe now that genetic engineers have found ways of reproducing humans non-sexually there are lessons in survival that we could learn from the plant kingdom.

34 Artificial Insemination

ARTIFICIAL INSEMINATION always sounds so much more like some dreadful weapon of war than a method of fecundation; more like something out of some biological time-bomb laboratory than a means of transferring semen from a donor to a recipient. It sounds as though it should only be used on cows, yet in point of fact man uses it not merely on these ungainly ungulates, but also among the creatures he holds most dear to his aesthetic sense and indeed, his heart, for he does it also to his dogs, his horses, and his women.

He can also do it with flowers, in order to obtain certain quite specific advantages. All that he has to do is perform a simple little operation on the flowers involved, an operation known as emasculation. What is actually involved is firstly the selection of plants or flowers that have the desired qualities. Then the male organs are carefully removed from the female. Each flower is then hermetically sealed within a hood of some sort in order to prevent pollination by natural agents. When the female organs are ready, pollen is removed from the anthers of the male flowers on the tip of a camel-hair brush and transferred to the tip of the stigma. It is in this way that artificial insemination is carried out on flowers.

The advantage of artificial insemination in cattle or horses is that a single outstanding stallion or bull can sire quite literally thousands of offspring in a single season, instead of the limited number he could sire if left to do it naturally. Further, the desirable genes can spread through a population of cattle or racehorses many times faster than could ever happen in the wild. We should be quite clear in our own minds that when we use a term such as 'advantages' we are making a qualitative judgement, and that that judgement is a wholly subjective one, made from man's point of view. It by no means follows from this that the changes made by man would be regarded as advantages by cows or racehorses.

In defence of artificial insemination it has been said that it accomplishes in a few decades what nature would have taken millions of years to achieve. The argument is a spurious one, since it misses the main point. Which is that what man selects in animals or plants as being desirable, is desirable from man's point of view: the qualities which would have been developed by the normal evolutionary methods are those which best fit a species for survival. It could well be that a cow with a three metre back and an udder capacity of fifteen litres is not what would have evolved but for man's interference, and maybe suits it least well for survival.

The same is true of the results achieved by man in the plant world. Artificial insemination in plants can achieve amazing results when judged from man's point of view. Thus pearl millet, which is sometimes known as poor man's rice, is grown extensively in some of the poorer regions of India, yet its yield has always been lower than that of rice. Within three years of the initiation of a breeding programme a hybrid was obtained which increased the edible yield by eighty-eight per cent thus more than fulfilling the aims of the programme.

Not all results are so spectacular, but there have been many outstanding successes. Deliberate, man-made hybrids between the European larch and the Japanese larch have produced plants which are bulking up to 180 per cent more than their parents. There is a man-made hybrid pea, of complex parentage, bred specially for today's harvesting, packing and marketing technology. It has virtually no leaves, which not only allows the sun to reach all the pods more evenly and so ensure their simultaneous ripening, it makes it easy to pick the pods by machine. Another success in the man-made hybrid story is that in some fodder it has been possible to increase the dry matter yield by as much as one-third.

Which is all very well for mankind, for it is man who benefits from these advances. It seems unlikely that man is bestowing any advantages on the plants. The probabilities are that left to their own devices the plants would not have chosen the path that man has chosen for them. For example, wild lupins are fast-growing legumes with lush foliage: they seem ideal as fodder. The trouble is that they contain an alkaloid that makes them most distasteful to cattle. It was argued that once in a while a plant would turn up free of this alkaloid. After searching through 1,500,000 plants man came up with a mere six plants in which this alkaloid was missing. Which means, if the sample was typical, that the chances of the alkaloid-free plant cropping up are 1,499,994 to 6 against. And the chances of any of those six plants growing sufficiently close to any other one of those six for them to cross pollinate each other are negligible. So the chances of an alkaloid-free strain originating in the wild are in the order of the wildest improbability: the genes are plainly there that enable it to happen in case it should need to happen in order to survive: but then so are thousands of genetic alternatives to the norm. Yet man, by selecting those six plants and breeding from them has produced a pure line of alkaloid-free lupins.

Man has made many attempts along these lines with different groups of plants in order to obtain plants which are, from his point of view, 'better', but in many instances the plants have irrefutably shown that they do not agree by being totally sterile

But the artificial insemination of plants conceals the dangers it is breeding. Because those plants in which man has most intensively selected for subjection to artificial insemination are those plants which form his staple diet – the grain crops – wheat, the staff of life, and rice, which in human terms actually supports more people than wheat. It has been possible to produce rices with three or four-fold yields over the last quarter-century, and wheats with similar increases in their yields. And the tendency where these high-yield plants exist, is to grow them to the total neglect of older varieties. There are some 3,000,000 acres of rice in the world, most of it man-made high-yield hybrids of very recent origin. The danger is not only that, in having bred the high-yield crop man may have destroyed the one gene which would safeguard the plant against a pest which could wipe out the whole 3,000,000 acres of modern hybrids, but that along the route he has discarded the diversity of genetic material in the older strains and wild grain crops. He could well find that by the time he has bred not only a rice that yields four times as much edible material, but also yields it four times a year instead of only twice a year, he has discarded and lost for ever the genetic variability which natural selection would have retained.

That in nature, natural organisms much prefer having their sexual relations in their own way, at their own time, should by now be apparent. Certainly in some areas nature is striking back, and is already ahead of man's technological grasp. There are primitive plants like bacteria which can feed on iron or attack concrete with sulphuric acid, and fungi which have adapted

themselves to living in aviation fuel, on rubber and even on some of the less durable plastics.

Artificial insemination may have assured us of jam today, but unless man tempers technology with wisdom he may not even have food tomorrow.

Index